To Oswald T. Avery

THE MYSTERIOUS

HUMAN GENOME

WORLD OF THE

FRANK RYAN

WILLIAM COLLINS

William Collins
An imprint of HarperCollins*Publishers*
1 London Bridge Street
London SE1 9GF

WilliamCollinsBooks.com

This William Collins paperback edition published in 2016

21 20 19 18 17 16
10 9 8 7 6 5 4 3 2 1

First published in Great Britain by William Collins in 2015

Text © FPR-Books Ltd 2015
Diagrams by Mark Salwowski

The author asserts the moral right to
be identified as the author of this work.

A catalogue record for this book
is available from the British Library.

ISBN 978-0-00-754908-5

Typeset by Palimpsest Book Production Limited, Falkirk, Stirlingshire
Printed and bound in Great Britain by
Clays Ltd, St Ives plc.

MIX
Paper from
responsible sources
FSC
www.fsc.org
FSC C007454

Possibly I am a scientist because I was curious when I was young. I can remember being ten, eleven, twelve years old and asking, 'Now why is that? Why do I see such a peculiar phenomenon? I would like to understand that.'

LINUS PAULING

contents

Introduction

> *No special act of creation, no spark of life was needed to turn dead matter into living things. The same atoms compose them both, arranged only in a different architecture.*
>
> JACOB BRONOWSKI, *THE IDENTITY OF MAN*

Bronowski begins his more famous book, *The Ascent of Man*, with the words, 'Man is a singular creature. He has a set of gifts which make him unique among the animals: so that, unlike them, he is not a figure in the landscape – he is a shaper of the landscape.' But why should we humans have become shapers of the landscape rather than mere figures inhabiting it? We differ from, say, a sea horse or a cheetah because our genetic inheritance, the sum of the DNA that codes for us, is different in humans when compared to the horse or the cheetah. We call this the genome, or, to be more specific in our case, the human genome.

Our genome defines us at the most profound level. That same genome is present in every one of the approximately 100,000 billion cells that make us who we are as individual members of the human species. But it runs deeper than that. In more personal terms, in myriad tiny variations that we each possess and are individual to us only, it is the very essence of us, all that, in genetic and hereditary terms, we have to contribute to our offspring, and through them to the sum total evolutionary inheritance of our species. To understand it is to know, in the most intimate sense, what it means to be human. No two people in the world today have exactly the same genome. Even identical twins, who will have been conceived with exactly the same genome, will have developed tiny differences between their genomes by the time of their births: differences that may have arisen in parts of their genomes that don't actually code for what we normally mean by genes.

How strange to realise that there is actually more to our individual genome than genes alone. But let us put aside such details for the moment to focus on the more general theme. How could a relatively simple chemical code give rise to the complexity of a human being? How could our human genome have evolved? How does it actually work? Immediately we are confronted by mysteries.

To answer these questions we need to explore the genome's basic structure, its operating systems and its mechanisms of expression and control. Some readers might react with incredulity. Surely any such exploration promises a journey into extraordinary complexity, one that is far too obscure and scientific for a non-scientist reader? In fact this book is aimed at exactly such a reader. As we shall see, the basic facts are easy enough to grasp, and the way to grasp them is to break the exploration into a series of simple, and eminently logical, stages. This journey will lead us through a sequence of remarkable revelations about our human history – even into the very distant past of our ancestors' lives and their prehistoric exploration of our beautiful life-giving planet.

The exploration will raise other, equally important, questions, too. How, for example, does this extraordinary entity that we call the human genome enable our human reproduction – the fertilisation of our maternal egg with the paternal sperm? How does it control the quasi-miracle of the developing embryo within the mother's womb? Returning to generalities for the moment, though, we can be sure that an important ingredient of the genome, and its essential nature, is memory – the memory, for example, of the totality of every individual human's genetic inheritance. But how exactly does it perform this remarkable feat of memory? We know already that this wonder chemical we call DNA works like a code, but how could any code recall the complex instructions that go into the making of cells and tissues and organs, and then once made, bring them into function in the single coordinated whole that comprises the human being? Even then we have hardly begun to confront the mysteries of the human genome. How does this same extraordinary structure acquire the programming that gifts the growing child with the wonder of speech, that bestows the related capacity for learning, writing and education, and which makes possible the maturing of the newborn to the future adult, who then repeats the cycle all over again when he or she becomes a parent in turn?

The wonder is that all of this might be encompassed in a minuscule cluster of chemicals including, but not exclusive to, the master molecule we call deoxyribonucleic acid – or DNA. This chemical code somehow records the genetic instructions for making us. Built into that code must be the potential for individual liberty of thought and inventiveness, enabling every human artistic, mathematical and scientific creativity. It gives rise to what each of us thinks innately as our private inviolate 'self'. Somehow that same construction of 'self' made possible the genius of Mozart, Picasso, Newton and Einstein. It is little wonder that we look at the repository of such potential with awe. No more is it surprising that we should want to understand this mystery that lies at the very core of our being.

Only recently have we come to understand the human genome in sufficient depth and subtlety to be able to put together its marvellous story – to discover, for example, that there is rather more to it than DNA alone. This is the story I shall attempt to convey in this book.

A few years ago I gave a lecture on a related theme at King's College London. The chairman asked me if I planned to write a book about it. When I said yes, he asked me to please write it in words that a lay reader, like himself, could readily understand.

'Just how simple do you want me to make it?'

'I want you to assume that I – your reader – know nothing at all to begin with.'

This, then, I promise to do. There will be no complicated scientific language, no mathematical or chemical formulae or unexplained jargon, and I shall introduce no more than a handful of simple illustrations. Instead I shall begin from first principles and assume that the readers of this book know little about biology or genetics. Even non-biologists might recall the many surprises when the first rough draft of the human genome was announced to the world, in 2001. The discoveries since then have confirmed that a good deal of the human genome – in its evolution, structure and workings – is far from what we had earlier imagined. Those surprises do not diminish the importance of the wealth of knowledge that had gone before, but rather, like all great scientific discoveries, they enhance it. Thanks to this new understanding, humanity has entered what I believe to be a golden age of genetic and genomic enlightenment, which is already being extrapolated to many important fields, from medicine to our human prehistory. I think society at large deserves to understand this and what it promises for the future.

one

Who Could Have Guessed It?

The large and important and very much discussed question is: How can the events in space and time which take place within the spatial boundary of a living organism be accounted for by physics and chemistry?

ERWIN SCHRÖDINGER

In April 1927 a young Frenchman, René Jules Dubos, arrived at the Rockefeller Institute for Medical Research, in New York, on what would appear to have been a hopeless mission. Tall, bespectacled and a recent graduate of Rutgers University, New Jersey, with a PhD in soil microbiology, Dubos had an unusually philosophical attitude to science. He had become convinced, through the work of eminent Russian soil microbiologist Sergei Winogradsky, that it was a waste of time studying bacteria in test tubes and laboratory cultures. Dubos believed that if we really wanted to understand bacteria we should go out and study them where they actually lived and interacted with one another and with life in general, in the fields and the woods – in nature.

On graduation from Rutgers Dubos had found himself unemployed. He had applied to the National Research Council Fellowship for a research grant but had been turned down because he wasn't American, but somebody had scribbled a handwritten message on the margin of the rejection letter. Dubos would later reflect upon the fact that it was written in a female hand, almost certainly added as a kindly afterthought by the official's secretary. 'Why don't you go and ask advice and help from your famous fellow countryman, Dr Alexis Carrel, at the Rockefeller Institute?' Dubos duly wrote to Carrel, which brought him, in April 1927, to the building on York Avenue, on the bank of the East River.

Dubos knew nothing about Carrel, or indeed about the Rockefeller Institute for Medical Research, and was surprised on his arrival to discover that Carrel was a vascular surgeon. Dubos had no academic knowledge of medicine and Carrel knew nothing about the microbes that lived in soil. The outcome of their conversation was all too predictable; Carrel was unable to help the youthful microbiologist. Their conversation closed about lunchtime and Carrel did Dubos the courtesy of inviting him to have lunch with him in the Institute's dining room, which had the attraction for a hungry Frenchman that they served freshly baked bread.

It seemed entirely by accident that Dubos found himself sitting at a table next to a small, slightly built gentleman with a domed bald head who addressed him politely in a Canadian accent. The gentleman's name was Oswald Theodore Avery. Although Dubos later confessed that he knew as little about Avery as he did of Carrel, Professor Avery (his close associates referred to him as 'Fess') was eminent in his field, which was medical microbiology. It would prove to be a meeting of historic importance both to biology and to medicine.

Avery subsequently hired Dubos as a research assistant, in which role Dubos discovered the first soil-derived antibiotics. Meanwhile

Avery led his small team – who were working on what he called his 'little kitchen chemistry' – on another quite different quest, one that would help unravel the key to heredity. Why then does society know little to nothing about this visionary scientist? To understand how this anomaly came about we need to go back in time to the man himself and the problems he faced three-quarters of a century ago.

In 1927, when Dubos first met Avery, the principles of heredity were poorly understood. The term 'gene' had been introduced into the nomenclature two decades earlier by a Danish geneticist, Wilhelm Johannsen. Curiously, Johannsen had adopted a vague concept of heredity, known as 'pangene', that was first proposed by Charles Darwin himself. Johannsen modified it to take on board the belated discovery of the pioneering work of Gregor Mendel, which dated back to the nineteenth century.

Readers may be familiar with the story of Mendel – the cigar-smoking, Friar-Tuck-like abbot of an Augustinian monastery in Brünn, Moravia (now the Czech Republic) – who undertook some brilliantly original studies of the peas he cross-bred in the monastery vegetable garden. From these studies Mendel discovered the basis of what we now know as the laws of heredity. He found that certain characteristics of the peas were transmitted to the offspring in a predictable manner. These characteristics included tallness or dwarfishness, presence or absence of the colours yellow or green in the blossoms or axils of the leaves, and the wrinkled or smooth skin in the peas. Mendel's breakthrough was to realise that heredity was stored within the germ cells of plants – and this would subsequently be extrapolated to all living organisms – in the form of discrete packets of information that somehow coded for specific physical characters or 'traits'. Johannsen coined the term 'gene' for Mendel's packet of hereditary information. At much the same time, a combative British researcher, William Bateson, extrapolated the term 'gene' to the discipline he now called 'genetics' to cover the study of the nature and workings of heredity.

Today, if we visit the free dictionary online, we get the following definition of a gene: 'The basic physical unit of heredity; a linear sequence of nucleotides along a segment of DNA that provides the coded instructions for the synthesis of RNA, which, when translated into protein, leads to the expression of hereditary character.' But Mendel had no notion of genes as such, and he certainly knew nothing about DNA. His discoveries languished in some little-read papers for forty years before they were rediscovered and their importance was more fully understood. But in time his idea about the discrete packets of heredity we now call genes helped to answer a very important medical mystery: how certain diseases come about through an aberration of heredity.

We now know that genes are the basic building blocks of heredity in much the same way that atoms are the physical units that make up the physical world. But during the early decades of the twentieth century, nobody had any real notion of what genes were made from, or how they worked. But here and there, people began to study genes in more depth by examining their physical expression during the formation of embryos or in the causation of hereditary diseases. Fruit flies became the experimental model for pioneering research in the laboratory of Chicago-based geneticist Thomas Hunt Morgan, where researchers located genes, one by one, onto structures called chromosomes, which were themselves located within the nuclei of the insect's germ cells. The botanical geneticist Barbara McClintock confirmed that this was also the case for plants. McClintock developed techniques that allowed biologists to visualise the actual chromosomes in maize, leading to the groundbreaking discovery that, during the formation of the male and female germ cells, the matching, or 'homologous', chromosomes from the two parents lined up opposite each other and then the chromosomes swapped similar bits so that the offspring inherited a jumbled-up mixture of the inheritance from the two parents. This curious genetic behaviour (which is known as 'homologous sexual recombination') is the explanation of why siblings are different from one another.

By the early 1930s biologists and medical researchers knew that genes were actual physical entities – chemical blocks of information that were lined up like beads in a necklace along the lengths of chromosomes. In other words, the genome could be loosely compared to a library of chemical information in which the books were the chromosomes. The discrete entities known as genes could then be compared to discrete words in the books. The libraries were housed in the nuclei of the germ cells – in human terms, the ova and sperm. Humans had a total stack of 46 books, which were the summed complement of ova and sperm, in every living cell. This came about because the germ cells – the ovum and the sperm – contained 23 chromosomes, so that when a human baby was conceived the two sets of the parental chromosomes united within the fertilised ovum, passing on the full complement of 46 chromosomes to the offspring. But this initial unravelling of the 'heredity mystery' merely opened up a Pandora's box of new mysteries when it came to applying genetics to the huge diversity of life on our fecund planet.

For example, did every life form, from worms to eagles, from the protozoa that crawled about in the scum of ponds to humanity itself, carry the same kinds of genes in their nuclei-bound chromosomes?

The microscopic cellular life forms, including bacteria and archaea, do not store their heredity in a nucleus. These are called the 'prokaryotes', which means pre-nucleates. All other life forms store their heredity in nuclei and are known collectively as 'eukaryotes', which means true nucleates. From the growing discoveries in fruit flies, plants and medical sciences, it was becoming rather likely – excitingly so – that some profound commonalities might be found in all nucleated life forms. But did the same genetic concepts, such as genes, apply to the prokaryotes, which reproduced asexually by budding, without the need for germ cells? At this time within the world of early bacteriology there was even a debate as to whether bacteria should be seen as life forms at all. And viruses, which were for the most part several orders of magnitude smaller than bacteria, were little understood.

Over time many researchers came to see bacteria as living organisms, classifying them according to the binomial Linnaean system; so, for example, the tuberculosis germ was labelled *Mycobacterium tuberculosis* and the boil-causing coccoid germ was labelled *Staphylococcus aureus*. Oswald Avery, with his extremely conservative nature, kept his options open, eschewing the binomial system and referring to the TB germ as the 'tubercle bacillus'. It is instructive for our story that Dubos, who came to know Avery better than any other colleague, would observe that 'Fess' was similarly conservative in his approach to laboratory research. Science must adhere with a puritanical stringency to what can be logically observed and definitively proven in the laboratory.

In 1882 German physician Robert Koch discovered that *Mycobacterium tuberculosis* was the cause of the greatest infectious killer in human history – tuberculosis. Koch constructed a code of logic that would be applied to bugs when first determining if they caused specific diseases. Known as 'Koch's postulates', this was universally adhered to, and once a causative bug had been identified it was studied further under the microscope. Thus the bug was duly classified in a number of ways. If its cells were rounded in shape it was a 'coccus', if a sausage shape it was a 'bacillus', if a spiral shape it was a 'spirochaete'. Bacteriologists methodically studied the sort of culture media in which a bug would grow best – whether in agar alone, or agar with added ox blood, and so on. They also studied the appearance of the bacterial colonies when they were grown in culture plates – their colours, the size of the colonies, whether they were rough in outline or round and smooth, raised or flat, stellate, granular or daisy-head. So the textbooks of bacteriology extended their knowledge base on a foundation of precise factual study and observation. And as understanding grew, this newfound knowledge was applied to the war against infection.

One of the useful things they learnt about disease-causing, or 'pathogenic', bacteria was that the behaviour of the disease, and thus of the bug itself in relation to its infected host, could be altered by various deliberate means: for example, through repeated cultures in the laboratory, or by repeatedly passing generations of the bug through a series of experimental animals. Through such manipulations it was possible to make the disease worse or less severe by making the bug either 'more virulent' or 'attenuated'. Bacteriologists looked for ways to extrapolate this to medicine. In France, for example, the eminent Louis Pasteur used this principle of attenuation to develop the first vaccine to be used successfully against the otherwise universally fatal virus infection of rabies.

One fascinating observation that came out of these studies was the fact that, once a bug had been attenuated or been driven to greater virulence, the change in behaviour could be 'passed on' to future generations. Could it be that some factor of the bug's own heredity had been altered to explain the change in behaviour?

Bacteriologists talked about 'adaptation', using the same term that was coming into vogue with evolutionary biologists when referring to evolutionary change in living organisms as they adapted

11

to their ecology over time. While it was too early to be sure if bacterial heredity depended on genes, these scientists linked it to the physical appearance of bugs and colonies, or to the bugs' internal chemistry, and even to their behaviour in relation to their hosts. These were measurable properties, the bacterial equivalents of what evolutionary biologists were calling the 'phenotype' – the physical make-up of an organism as opposed to what was determined by the hereditary make-up, or 'genotype'.

Bacteriologists also came to recognise that the same bacterium could exist in different subtypes, which could often be distinguished from one another using antibodies. These subtypes were called 'serotypes'. In 1921 a British bacteriologist, J. A. Arkwright, noticed that the colonies of a virulent type of dysentery bug, called *Shigella*, growing on the jelly-coated surfaces of culture plates, were dome-shaped with a smooth surface, whereas colonies of an attenuated, non-virulent, type of dysentery bug were irregular, rough-looking and much flatter. He introduced the terms 'Smooth' and 'Rough' (abbreviated to S and R) to describe these colonial characteristics. Arkwright recognised that the 'R' forms cropped up in cultures grown under artificial conditions, but not in circumstances where bacteria were taken from infected human tissues. He concluded that what he was observing was a form of Darwinian evolution at work.

In his words: 'The human body infected with dysentery may be considered a selective environment which keeps such pathogenic bacteria in the forms in which they are usually encountered.'

Soon researchers in different countries confirmed that loss of virulence in a number of pathogenic bacteria was accompanied by the same change in colony appearance from Smooth to Rough. In 1923, Frederick Griffith, an epidemiologist working for the Ministry of Health in London, reported that pneumococci – the bugs that caused epidemic pneumonia and meningitis which were of particular interest to Oswald Avery at the Rockefeller Laboratory – formed similar patterns of S and R forms on culture plates. Griffith was known to be a diligent scientist and Avery was naturally intrigued.

Griffith's experiments also produced an additional finding, one that really shook and puzzled Avery.

When Griffith injected non-virulent R-type pneumococci from the strain known as type I into experimental mice, he included an additional ingredient in the injections, a so-called 'adjuvant', which usually pepped up the immune response to the R pneumococci. A common adjuvant for these purposes was mucus taken from the lining of the experimental animal stomach. But for some obscure reason Griffith switched adjuvant to a suspension of S pneumococci, derived from type II, that had been deliberately killed off by heat. The experimental mice died from overwhelming infection. In the blood of these dead mice Griffith expected to find large numbers of multiplying R-type I bacteria – the type that he had injected at the start of the experiment. Why then had he actually found S-type II? How on earth could adding dead bacteria to his inoculum have changed the actual serotype of the bacterium from non-virulent R-type I to highly virulent S-type II?

Researchers, including Avery himself, had previously shown that S and R types were determined by differences in the polysaccharide capsules coating the cell bodies of the bugs. Griffith's findings suggested that the test bacteria, initially R-type pneumococci, had changed their polysaccharide coats inside the infected bodies of the mice to that of the virulent strain. But they could not have achieved this by just flinging off the old coat and putting on the new one. The coat was determined by the bacteria's heredity – it was an inherited characteristic. Further cultures of the recovered bacteria confirmed that the S type bred true. There appeared to be only one possible explanation: adding the dead S bacteria to the living R bacteria had induced a mutation in the heredity of the living R-type bacteria, so they literally transformed into S-type II.

In the words of Dubos: '[At the time] Griffith took it for granted that the changes remained within the limits of the species. He probably had not envisaged that one pneumococcus type could be transformed into another, as this was then regarded as the equivalent of transforming one species into another – a phenomenon never previously observed.'

*

It is little wonder that Avery was astonished by Griffith's findings. Like Robert Koch before him, Avery subscribed to the view that bacterial strains were immutable in terms of their heredity. The very concept of a mutation – that heredity was capable of an experimentally induced change – was a highly controversial issue within biology and medicine at this time. To understand why, we need to grasp the concept of what a mutation means.

By the late nineteenth century Darwinian theory had entered a crisis. Darwin himself had been well aware that natural selection relied on some additional mechanism, or mechanisms, capable of changing heredity, so that natural selection would have a range of 'hereditable variation' to choose between. Generations later, in the opening chapters of his innovative book *Evolution: The Modern Synthesis*, Julian Huxley put his finger on the nub of the problem. 'The really important criticisms have fallen upon Natural Selection as an evolutionary principle and centred round the nature of inheritable variation.' In 1900, a Dutch biologist, Hugo de Vries, put forward a novel mechanism that would be capable of providing the necessary variation: the concept of a random change in a unit of inheritance. Opportunity for change exists when genes are copied during reproduction, when a random change in the coding of a gene might arise from an error in copying the hereditary information. De Vries called this source of hereditary change a 'mutation'. It was only with what Julian Huxley termed 'the synthesis' of Mendelian genetics – the potential for change in the inherited genes through mutation – and Darwinian natural selection operating on the hereditary choices presented within a species, that Darwinian theory became credible again to the great majority of scientists.

In time Griffith's finding would be confirmed to be what Avery was now wondering about: it was a mutation. Geneticists would show that the change from the R to the S strain of pneumococcus involved the transfer of a gene from the dead S-type II bacteria to the living R-type I bacteria, which was incorporated into subsequent bacterial reproductive cycles, transforming the cells of the R-type I bacterium into the cells of the S-type II bacterium. It was indeed the bacterial equivalent of a change of species. And Griffith was proven right in inferring that Darwinian natural selection had operated even in the short time frame of the infection of a cohort of laboratory mice.

Griffith's experimental findings galvanised bacteriologists and immunologists around the world. His discovery was confirmed in several different research centres, including the Robert Koch Institute in Berlin, where the pneumococcal types had first been classified. The news was inevitably a hot topic of discussion in Avery's department, as Dubos would recount: 'but we did not even try to repeat them at first, as if we had been stunned and almost paralysed intellectually by the shocking nature of the findings'.

At first Avery simply couldn't believe that bacterial types could be transformed. Indeed, he had been one of the authoritative figures who had settled the fixity of bacterial reproduction being true to type years before. But from 1926 Avery encouraged a young Canadian physician working in the Rockefeller Laboratory,

M. H. Dawson, to investigate the situation. According to Dubos, Dawson, unlike Avery, was convinced from the start that Griffith's conclusion must be correct because he believed that 'work done in the British Ministry of Health had to be right'.

Dawson began by confirming Griffith's findings in laboratory mice. His results suggested that the majority of non-virulent bacteria – the R types – had the ability in certain circumstances to revert to the virulent S type. By 1930 the young Canadian was joined by a Chinese colleague, Richard P. Sia, and between them they took the experimental observations further by confirming that the hereditary transformation could be brought about in culture media, without the need for passage through mice. At this stage, Dawson left the department and Avery encouraged another young physician, J. L. Alloway, to take the investigation further. Alloway discovered that all he needed to bring about the transformation was a soluble fraction derived from the S pneumococci by dissolving the living cells in sodium deoxycholate, then passing the resultant solution through filters to remove the bits of broken-up cells. When he added alcohol to the filtered solution, the active material precipitated out as sticky syrup. Throughout the laboratory this sticky syrup was referred to as the 'transforming principle'. So the work continued, experiment following experiment, year by year.

When Alloway left the department, in 1932, Avery began to devote some of his own time to the pneumococcus transformation, in particular aiming to improve the extraction and preparation of the transforming substance. Frustration followed frustration. He focused on its chemical nature. Discussion took place with other members of the department, ranging from the 'plamagene' that was thought to induce cancer in chickens (now known to be a retrovirus), or to the genetic alterations in bacteria that were thought to be caused by viruses. According to Dubos, Alloway suggested the transforming agent might be a protein-polysaccharide complex. But by 1935 Avery was beginning to think along other lines. In his annual departmental report that year he indicated that he had obtained the transforming material in a form that was essentially clear of any capsular polysaccharide. In 1936, Rollin Hotchkiss, a biochemist who had now arrived to work in the department, wrote a historic comment in his personal notes:

'Avery outlined to me that the transforming agent could hardly be a carbohydrate, did not match very well with a protein and wistfully suggested it might be a nucleic acid!' At this stage, Dubos, who many years later would write a book about Avery and his work, dismissed this as no more than a surmise. There were good reasons for his caution.

That year few researchers throughout the world believed that the answer to heredity lay with nucleic acids. These chemical entities had been discovered by a Swiss biochemist, Johann Friedrich Miescher, back in the late 1800s. Fascinated by the chemistry of the nucleus, Miescher had broken open the nuclei of white blood cells in pus, and subsequently the heads of salmon sperm, to discover a new chemical compound which was acidic to pH testing, rich in phosphorus and comprised of enormously large molecules. After a lifetime of experimentation on the discovery, Miescher's pupil, Richard Altmann, would introduce the term nucleic acid to describe Miescher's discovery. By the 1920s, biochemists and geneticists were aware that there were two kinds of nucleic acids. One was called ribonucleic acid, or RNA, which contained four structural chemicals: guanine, adenine, cytosine and uracil, or GACU. The other was called 'desoxyribonucleic acid', or DNA, which was a major component of the chromosomes. They had deciphered its four bases – three identical to RNA, guanine, adenine and cytosine, but with the uracil replaced by thymine – making the acronym GACT. They knew that these four bases consisted of two different pairs of organic chemicals; adenine and guanine being purines, and cytosine and thymine being pyrimidines. They also knew that they were strung together to form very long molecules. At first they thought that RNA was confined to plants while DNA was confined to animals, but by the early thirties this was dismissed when it was found that both RNA and DNA were universally distributed throughout the animal and plant kingdoms. Still they had no knowledge of what nucleic acids actually did in the nuclei of cells.

A distinguished organic chemist based at the Rockefeller Institute, Phoebus Aaron Levene, proposed that the structures of DNA and RNA were exceedingly boring – they formed groups of four bases that repeated themselves in the identical repetitive formation throughout the molecule, like a four-letter word, repeated ad nauseam. This was called 'the tetranucleotide hypothesis'. Such a banal molecule couldn't possibly underlie the exceedingly complex basis of heredity. In the words of Horace Freeland Judson, 'the belief was held with dogmatic tenacity that DNA could only be some sort of structural stiffening, the laundry cardboard in the shirt, the wooden stretcher behind the Rembrandt, since the genetic material would have to be protein'.

Proteins are lengthy molecules made up of smaller organic chemical units known as amino acids. There are 20 amino acids in the make-up of proteins, reminiscent of the number of letters that make up alphabets. If genes were the hereditary equivalents of words, only the complexity of proteins could fashion the words capable of spelling out the narratives. Chemists, and through extrapolation geneticists, not unnaturally assumed that only this level of complexity could possibly accommodate the incredible memory template that the complexity of heredity demanded – a line of thought that Judson labelled 'The Protein Version of the Central Dogma'.

This was the contentious zeitgeist that Avery now confronted. As early as 1935, in his annual reports to the Board of the Institute, he indicated that he had growing evidence that the 'transforming substance' appeared free of capsular polysaccharide and it did not appear to be a protein.

Further progress on this line of research appeared to drag. In part this was because Dubos, working in the same department, had made a breakthrough in his search for antibiotic drugs. In 1925, Alexander Fleming, at St Mary's Hospital in London, had discovered a potential antibiotic, penicillin, but he had been unable to take his work to the stage of useful production for medical purposes. Now, working on the philosophical principle encapsulated by the biblical saying 'dust to dust', Dubos had pioneered the search for microbes in soil that would potentially attack the polysaccharide coat of the pneumococcus. By the early 1930s he was making progress. From a cranberry bog in New Jersey he found a bacillus that dissolved the thick polysaccharide capsule that coated the pneumococcus with its armour-like outer covering. Dubos went on to extract the enzyme that the Cranberry Bog bacillus produced. He and Avery had reported their discovery in a paper in the journal, *Science*, in 1930. In a further series of papers the two scientists would report further experiments, all aimed at extrapolating the discovery to human trials of the Cranberry Bog enzyme in treating the potentially fatal pneumonia and meningitis caused by the pneumococcus.

But their researches encountered difficulty after difficulty. In part these arose from a predictable ignorance in a field of such pioneering research. A more personal, and devastating, problem arose when, under the stress of it all, Avery developed thyrotoxicosis – a debilitating autoimmune illness in which his thyroid gland became overactive.

Thyrotoxicosis causes the system to be flooded by thyroid

hormones, which would have inappropriately switched his metabolism into a dangerous overdrive. He would have felt shaky, agitated, physically and mentally restless, suffering difficulties with relaxation and sleep – an impossible situation for a creative person. Avery had to spend time away from the lab undergoing surgery to remove the bulk of the 'toxic goitre', a procedure that carried risk of side-effects, even fatality in a minority of cases. His surgeon advised him against any early activity, physical or mental, that provoked stress. Dubos later recalled how Avery was away from his work for as long as six months. And while Avery was away, the laboratory stagnated. In Dubos' own words, 'I . . . pursued [the research] for three or four years. However I could not carry the work very far because there were serious gaps in both my knowledge of genetics and biochemistry and in the [prevailing] states of these sciences themselves.'

Dubos would continue his researches against such difficulties, to be rewarded, in 1939, with the discovery of the first soil-derived antibiotic. He called it 'gramicidin'. But gramicidin could not be taken by mouth or administered by injection because it was too toxic. It could only be applied to skin conditions. The research continued. But then, all of a sudden, the hopes of Avery and Dubos were overtaken by a rival breakthrough. Working in the pharmaceutical research laboratories of the Bayer Company in Elberfeld, Germany, doctor Gerhard Domagk reported the discovery of a new antibacterial agent called prontosil. The first of what would come to be known as the sulphonamide drugs, it immediately entered the medical formulary, pioneering the treatment of a number of hitherto untreatable infectious diseases.

Today we are apt to forget how little we could do to control infection in the 1930s. Epidemics such as scarlet fever, measles, pneumonia, meningitis and poliomyelitis swept through the population in regular, sometimes annual, cycles. Other notorious infections were everyday threats, including tuberculosis, which ravaged entire families, or boils, septic arthritis, septic osteomyelitis, which caused agonising abscesses in bone, and the commonplace but potentially deadly streptococci capable of breaking through a septic throat to cause abscesses in the brain. Most of the human population, whether in developed or developing countries, died from infections, including the insidious pneumonias that hit those whose immunity was depressed. The treatment of infections was the most urgent problem then facing humanity. For Dubos, and even more so Avery, the disappointment of failing in their line of research would have been shattering.

When, in due course, Avery returned to work, he switched the emphasis of his research to the 'transforming substance'. Colin MacLeod improved the techniques of extraction so they could now produce sizeable amounts for assay and further testing. They began to make more rapid progress so that, in a report to the Rockefeller Board for the year 1940–41, they were more confident in stating that even a highly purified extract of the transforming substance appeared to be protein-free.

That summer MacLeod left the Institute to become Professor of Bacteriology at the New York University School of Medicine. But he still took an interest in the project and frequently returned to the Institute to add his advice. A young paediatrician, Maclyn McCarty, took MacLeod's place in the transforming experiment. McCarty brought a useful level of biochemical training to the laboratory. And now they had the transforming substance in quantity and in stable form, he applied his chemical skills to further process and identify the active material. He began to culture the pneumococci in large batches of 50 to 75 litres, developing a series of steps that increased the yield of transforming substance while removing proteins, polysaccharides and ribonucleic acid. The prevailing beliefs about the hereditary principle claimed that nucleoproteins were the answer. Thus the topmost priority in all of this effort was to ensure that the final test material contained no protein.

By now McCarty had extracted concentrated solutions of the active material. He treated this with a series of protein-digesting enzymes, such as the gut-derived trypsin and chymotrypsin, which were known to destroy proteins, ribonucleic acid and pneumococcal capsular polysaccharide. What remained was once more shaken with chloroform in a final effort to remove even the finest traces of protein.

By late 1942, after repeated extraction and experiment, McCarty had come to the conclusion that the transforming activity was confined to a highly viscous fraction that consisted almost exclusively of polymerised deoxyribonucleic acid. When he precipitated this fraction in a flask by adding absolute ethyl alcohol, drop by drop, all the while stirring the solution with a glass rod, the active material separated out of the solution in the form of long, white and extremely fine fibrous strands that wound themselves around the stirring rod. Dubos would recall the excitement felt within the lab by all those who witnessed the sight of the beautiful fibres, which were the pure form of the transforming substance.

In early 1943, Avery, MacLeod and McCarty presented their findings to distinguished chemists at the Princeton section of the Rockefeller Institute for Medical Research. The chemists must have been astonished, perhaps even nonplussed, but they offered no contradiction of the evidence nor asked for further proof. The researchers summed up the evidence for the Board of the Rockefeller in April of that year. Avery, MacLeod and McCarty, all three medical doctors rather than geneticists, were now ready to inform the world in a paper submitted to the *Journal of Experimental Medicine* in November the same year, which would be published early the following year. The title of the paper was long-winded and cautious: 'Studies on the chemical nature of the substance inducing transformation of pneumococcal types. Induction of transformation by a desoxyribonucleic acid fraction isolated from pneumococcus type III'.

In the words of Dubos, this paper 'had staggering implications'. The sense of excitement, tempered by caution, was captured in a letter that Avery wrote to his brother, Roy, dated 26 May 1943:

> . . . *For the past two years, first with MacLeod and now with Dr McCarty, I have been trying to find out what is the chemical nature of the substance in the bacterial extracts which induces this specific change . . . Some job – and full of heartaches and heartbreaks. But at last perhaps we have it . . . In short, the substance . . . conforms very closely to the theoretical values of pure desoxyribose nucleic acid. Who could have guessed it?*

In the letter, 'desoxyribose nucleic acid', in the paper, 'desoxyribonucleic acid': these are older names for what we now call deoxyribonucleic acid – commonly reduced to its acronym, DNA.

two

DNA Is Confirmed as the Code

Looking back at his own failure to appreciate Avery's discovery at the time, Stent came to the conclusion 'in some respects Avery et al's *paper is a more dramatic example of prematurity than Mendel's'.*

UTI DEICHMANN

Scientists, in the opinion of the Nobel Prize-winning Linus Pauling, were fortunate because their world was so much the richer for its mysteries than those not interested in science could possibly appreciate. Certainly in those days Avery's lab at the Rockefeller Medical Institute for Research was filled with a mood of expectation and excitement. In 1943 Oswald Avery was 65 years old. He had planned to retire and join his brother Roy's family in Nashville, Tennessee, but there was no question of his leaving the lab at this time. He needed to continue his work on the transforming substance. In particular he needed to convince his colleagues throughout the world of microbiology and, more widely, the sceptical world of biochemists and geneticists, of the validity of their discovery.

Avery was conservative by nature. A generation earlier he and a colleague had proposed that complex sugar molecules, called polysaccharides, and not proteins determined the immunological differences between different types of pneumococcal bacteria. Although this theory was eventually confirmed to be true, at the time of discovery it provoked a storm of controversy that had haunted this nervous and sensitive man. In a long and rambling letter to his brother Avery had repeatedly referred to his worry about the reaction to the new discovery. 'It's hazardous to go off half-cocked . . . It's lots of fun to blow bubbles – but it's wiser to prick them yourself before someone else tries to.'

Avery had an adversary closer to home. Alfred E. Mirsky, a distinguished biochemist and geneticist also working at the Rockefeller Institute, had reacted to Avery's discovery with incredulity. To make matters worse, Mirsky was widely regarded as an expert on DNA. He had discovered that the quantity of DNA in every cell nucleus remained the same, establishing a principle called 'DNA constancy'. He now doubted the efficacy of McCarty's DNA extraction. A stickler for 'clean' biochemical experiment, Mirsky believed that protein found in the nucleus, called nucleoprotein, must be the basis of heredity. Even as late as 1946, Mirsky insisted that the two enzymes McCarty had used in his extractions would not digest away all of the protein. Mirsky was very influential in genetic circles and his argument impressed the leading geneticist of the time, Hermann J. Muller, who had been awarded the Nobel Prize that same year for his discovery, made two decades earlier, that X-rays caused mutations in the genes of the fruit fly. In a letter to a geneticist colleague, Muller stated 'Avery's so-called nucleic acid is probably nucleoprotein after all, with the protein too tightly bound to be detected by ordinary method.'

To some extent such disagreement was typical of the situation one might find anywhere in science when various groups from different scientific backgrounds are investigating a major unknown. Never is the argument more acrimonious than when a new discovery confounds the accepted paradigm. But the vociferous opposition of Mirsky from within Avery's home research foundation must have been particularly damaging. In 1947 Muller published his 'Pilgrim's Lecture' as a scientific paper in which he concluded that whether nucleic acid or protein was the answer 'must as yet be regarded as an open question'. In the words of Robert Olby, a historian and philosopher of science, 'Through Muller's widely read Pilgrim Lecture, this [sceptical] influence was spread to a wide audience.'

In a new series of extractions, with stringent quality checking, Avery attempted to confound his critics. McCarty left the laboratory in 1946, which was left in the hands of, amongst others, the meticulous Rollin Hotchkiss. Hotchkiss added several new chemical explorations of the extract, all further confirming that it was DNA. He disproved Mirsky's objection by purifying the extract to the extent that the protein content was below 0.02 per cent and he showed that it was inactivated by a newly discovered crystalline enzyme specific to DNA: DNase. While many geneticists remained obdurate in their opposition, some were beginning to take notice.

In a subsequent interview with the biophysicist and future Nobel Laureate, the German-born physicist Max Delbrück, Horace F. Judson would discover that some distinguished researchers were aware of the potential importance of Avery's discovery. 'Certainly there was scepticism,' Delbrück recalled. 'Everybody who looked at it was confronted by this paradox. It was believed that DNA was a stupid substance . . . which couldn't do anything specific. So one of these premises had to be wrong. Either DNA was not a stupid molecule, or the thing that did the transformation was not DNA.' Avery had raised a monumentally important question and the only way of resolving the dilemma was for other researchers to probe it through some form of alternative experimentation to find out if he was right or wrong.

In 1951, two American microbiologists, Alfred Hershey and Martha Chase, undertook such an alternative experiment while studying the way that certain viruses use bacteria as a factory to make daughter viruses. These viruses are called 'bacteriophages', or 'phages' for short – from the Greek *phago*, which means to eat, because they devour cultures of host bacteria. The presence, and number, of viruses could be measured if you seeded your host bacteria into heat-softened agar and then added the viruses in various dilutions to the agar before spreading it over a laboratory plate. When the agar cooled it formed a thin, even layer of jelly in the plate, which, on overnight culture, would become clouded by growth of bacteria within the agar. Wherever a virus landed among the bacteria there would be a round window of transparency caused by the dissolving (lysis) of the bacteria which was easily visible, and thus countable. This 'plaque-counting technique', which I myself learnt from my microbiology professor as a medical student and later made use of in experiments on the nature of autoimmunity as a hospital doctor, is easily learnt and thus put to use by thousands of scientists in a great variety of experiments.

What interested Hershey and Chase was the fact that phage viruses were known to compose a core of genetic material surrounded by a capsule of protein. In fact, each virus closely resembled a medical syringe in structure, so that when it infected the bacterial cell of its host, it appeared to squeeze out the genetic material from the body of the syringe, leaving the empty protein coat attached to the outer bacterial cell wall. Meanwhile, the genetic material was injected into the bacterial cell interior, where the viral genome would be replicated as part of its reproduction.

Hershey and Chase invented an ingenious experiment that would decide whether protein or DNA was the basis of the viral reproductive system. This would involve adding radioactive phosphorus and radioactive sulphur to the media in which separate batches of the host bacteria were growing. After four hours, to allow the radioactive element to be taken up by the bacteria, they introduced the phage viruses.

To understand the basis of the experiment we need to grasp that DNA contains phosphorus as part of its make-up but no sulphur, meanwhile the amino acids that make up proteins contain sulphur but no phosphorus.

By inoculating each of these two groups of bacteria with viruses, Hershey and Chase derived two populations of phage viruses – one containing the radioactive phosphorus and the other containing the radioactive sulphur. When the viruses infected the bacteria, they left their empty viral coats, mostly made up of protein, attached to the outside of the bacterial cell walls, having injected their core material, known to comprise DNA, into the bacterial bodies. Hershey and Chase used centrifugation to separate and extract empty viral coats. Meanwhile, the infected bacteria were allowed to go through their normal reproductive cycle, which allowed the viral cores inside them to generate entire new phage viruses, rupturing the bacterial bodies and flooding the growth media with large numbers of fully formed viruses. Hershey and Chase now removed what was left of the host bacterial bodies to gather dense concentrations of fully formed viruses.

When they now compared the empty viral coats, made up of protein, with the fully formed viruses, with their cores full of genetic material, they found that 90 per cent of the radioactive sulphur was left behind in the viral coats when the virus infected the cell, and 85 per cent of the phosphorus was now part of DNA that had entered the bacterial cell to code for the future offspring of virus. This confirmed Avery's findings: DNA, and not protein, was the code of heredity.

We might duly note that this separation of coat from core DNA of virus involves a much higher degree of protein impurity than Avery's extractions. Yet the hitherto sceptical geneticists appeared to be more convinced by the phage experiment than by Avery's work. Perhaps the strikingly visual nature of the experiment was a factor. Perhaps it was the additional, quite different, avenue of confirmation. It didn't harm credibility that leading geneticists were within the 'phage camp', too.

*

Today, with the advantage of retrospect, scientists by and large see the 1944 paper by Avery, MacLeod and McCarty as the pioneering discovery of DNA as the molecule of heredity. It has been portrayed as one of the most regrettable examples of a discovery that merited, but was not awarded, the Nobel Prize. There is ample evidence that Avery was recommended by senior colleagues, particularly within his own discipline of microbiology and immunology – indeed he was nominated twice, first in the late 1930s, for his work on the pneumococcal typing and its relevance to bacterial classification, and, after the 1944 paper was published, he was nominated yet again for his fundamental contribution to biology. But it would appear that the Nobel Committee was not sufficiently swayed. In retrospect, it is seen as a major omission that causes people to scratch their heads and wonder why.

Dubos worked for fifteen years in the lab next door to Avery's and, in so much as the reticent Professor allowed it, he had plenty of opportunity to get to know Avery and to understand his approach to science and his reaction to the stresses involved in pioneering new concepts. In Dubos' opinion, writing in 1976, the curious lack of recognition most likely derived from a combination of happenstance and Avery's own personality. He would subsequently remark how, in all that time, Avery never closed the door of his lab, or the small office that led off it, allowing any of his staff to come and talk to him. This same eternally open door also allowed Dubos to witness 'Fess's' activities at the bench, to listen in to his conversations with colleagues and to observe his interludes of introspective brooding.

This reserved, small and slender bachelor would inevitably arrive at work dressed in a neat and subdued style, his conservative attire somehow at one with the charm of his lively and affable behaviour. His eyes, under the domed bald head that seemed too voluminous for the frail body, were sparkling and always questioning, and he would transform the most ordinary conversation into an artistic performance with spirited gestures, mimicry, pithy remarks and verbal pyrotechnics. Avery might have been somewhat reticent in manner (he could be silently introspective), but in his own quintessential way he was vulnerably human, and that made him all the more interesting and enchanting.

I would suggest that creativity in science is every bit as intertwined with personality as one finds in a writer, artist, or musically gifted composer or performer. It would seem unsurprising in an artist if he appeared unusually ascetic, withdrawn from the hurly-burly world of the surrounding New York, ensuring that he lived close enough to the Rockefeller Institute so he could walk to work. In his ways, Avery could seem curiously ambivalent. He suffered mood swings at times, when alone in the lab, when he would appear to be dejected by the difficulties facing him. Afterwards he would declaim, clearly referring to himself, that resentment hurts the person who resents much more than the person who is resented. He left many letters unanswered and refused to have a secretary. He refused to review, or sponsor, any scientific paper in which he had made no contribution. In Dubos' words, 'Graciousness and toughness when it came to what he himself was determined to do, was part of his nature.' Avery was a very successful teacher during his early medical career, yet in his later years he appears to have resented being expected to lecture on his own research. In this respect, he bore some interesting similarities to Charles Darwin. Avery scrupulously avoided any discussion of his own health and any intrusion, however small, into his private life – which was devoted to his younger brother, Roy, and to an orphaned first cousin whom he supported all through his life. He never expressed resentment about criticisms of his work, even when these were unjustified. He left no record of his private thoughts, other than the letters to his brother. A single experience struck Dubos as being significant.

One day, in early 1934, the same year that Avery suffered the onset of his thyrotoxicosis, Dubos told Avery that he was about to be married. The lady in question was a Frenchwoman living in New York, named Marie Louise Bonnet. Avery immediately rejoiced at the news. They were conversing in the laboratory on the sixth floor of the Rockefeller hospital building. During the subsequent animated conversation, Avery climbed out of his chair, walked to the window and looked out, as if lost for a moment in deep reflection. Returning to his chair, he mentioned that he had contemplated marriage years before, but that circumstances had not proved favourable to his plans. It seems likely that the lady in question was a nurse that Avery had met during the course he had taught to student nurses at the Hoagland Laboratory. Avery would have been about 32 years old at the time. For a moment or two the older man's eyes were full of longing.

'One of the great joys of life,' he remarked to Dubos, 'is to go home to someone who would rather see you than anybody else.'

Fate would prove cruel to both men. Marie Louise Bonnet

subsequently died from tuberculosis at a time when Dubos was pioneering the very antibiotics that would eventually help to cure the same illness. The marriage was childless and the effects of his wife's death on Dubos were devastating. He resigned, forthwith, from his antibiotic researches, which were later taken up by his former teacher, Selman Waksman at Rutgers Agricultural College, now Rutgers University, and which led to the discovery of a series of important antibiotics, including streptomycin. This break-through resulted in Waksman being awarded the Nobel Prize in Medicine or Physiology in 1952.

Much of what Dubos witnessed of Avery spoke of an intense focus and purity of purpose in science and his work. But, increasingly, his devotion to his research appeared to be accompanied by insularity bordering on reclusiveness.

Scientists who have laboured long and hard at a difficult but eventually rewarding line of research are usually happy to talk about it – if not to the media or ordinary social channels, certainly to colleagues. They travel to scientific symposia. They take part in conferences. They enjoy the camaraderie that comes from sharing the same interests. In the words of Frank Portugal, 'wide-ranging discussions with peers both individually and at meetings are part and parcel of the scientific process. It is an important component of how collaborations are formed and scientific advances are made and respected.' Most scientists are only too glad to accept the, often rare, honours and distinction their work brings their way. Not so Oswald Avery.

In 1944 Avery was proposed for an honorary degree at Cambridge University, a recognition most scientists would cherish. The following year he was awarded the Copley Medal by the Royal Society of London. Avery's roots were English – in the late nineteenth century his family had emigrated to Canada from the city of Norwich – but he refused to visit England even on such prestigious occasions, putting forward the excuse that his state of health did not permit it except by travelling first class. In Dubos' opinion, this was disingenuous, since the respective foundations would have funded the flights. That he might have felt nervous, claustrophobic, on the lengthy flight is possible. Recalling those dark moods in which Avery might mumble to himself about the damaging inflictions of resentment, it seemed more than likely to Dubos that he might have been unable to suppress lingering anger at the hurtful controversy provoked years ago by his polysaccharide typing of pneumococci. Whatever his reasons, Avery refused both honours.

An incident highlighted just how strong was Avery's aversion to such formal acknowledgement of his work. Sir Henry Dale, who was President of the Royal Society in England, took it upon himself to bring the Copley Medal to the Rockefeller Institute, there to confer it on the shy and retiring Avery in person. Dale was accompanied by a Dr Todd, who knew Avery personally. The two highly respected English visitors arrived at the Institute in New York unannounced and went directly to Avery's department in the main hospital building. But when they saw Avery working in his lab, through the ever-open door, they retreated without intruding on his presence.

Dr Todd would later recount how Sir Henry Dale said simply: 'Now I understand everything.'

Bizarre as this behaviour would appear, it was in keeping with Avery's increasingly reclusive personality: a man who avoided any of the normal personal contacts outside of immediate family and work colleagues. Genius can be strange. Yet such idiosyncratic behaviour apart, it was this son of an evangelical Baptist preacher who first discovered that DNA was the molecule of heredity. And putting such personal matters aside, the question remains: why was such a fundamental discovery not recognised by the awarding of the Nobel Prize?

In his letters to his brother, Avery retained a modest outlook. Could it be that a combination of Avery's innate conservatism, his tendency to over-caution, and his downplaying of the implications of his discovery in the paper of 1944 might have contributed to his being overlooked? In Dubos' words, the paper . . . 'did not make it obvious that the findings opened the door to a new era of biology'. Dubos wondered if the Nobel Committee, unaccustomed to such restraint and self-criticism 'bordering on the neurotic' might have caused them to wait a while for both confirmation of the discovery and to see what the broader implications might be. To put it another way, Dubos questioned if the paper might have been a failure not in its own merits, as a scientific communication, but from the public relations point of view.

This lack of recognition is made all the more puzzling by the fact that, if the importance of the 1944 paper was not universally recognised when it was published, it became more and more obvious with the passage of time. The Hershey and Chase paper was published in 1952. And although he was retired by the time Crick and Watson published their famous discovery of the three-dimensional chemical structure of DNA in 1953, Avery was still alive. He wouldn't die until two years later, in 1955.

More recently the Nobel authorities have allowed open access to their earlier thinking, and this has confirmed much of what Dubos had concluded. As part of the system for deciding who should get Nobel Prizes, the Nobel Committee receives proposals from leading experts around the world. In the words of Portugal, who reviewed their working and archives, 'It seems that key biological chemists were not convinced by Avery's claim that DNA was the basis of heredity.' Not a single geneticist nominated Avery for the Nobel Prize. In part this may have reflected a difficulty in extrapolating his discovery in a single type of bacterium to genetics more widely, but even those colleagues who did nominate him for the Nobel Prize tended to overlook his work on DNA in favour of his immunological typing of the pneumococcal capsule. Portugal also wondered if Avery's own idiosyncratic behaviour, including his reluctance to meet with and exchange findings with colleagues, and in particular geneticists, at scientific meetings had unintentionally confounded the acceptance of his groundbreaking discovery.

We are left with a lingering sense of regret that Avery was not accorded the recognition he deserved. He was 67 years old when his iconoclastic paper was published. It was, in the words of the eminent biochemist Erwin Chargaff, the rare instance of an old man making a major scientific discovery. 'He was a quiet man: and it would have honoured the world more, had it honoured him more.'

But there is a greater acknowledgement of discovery than the awarding of a prize, no matter how respected and prestigious. In the words of Freeland Judson, 'Avery opened up a new space in biologists' minds.' By space he meant he had unravelled a major truth, revealing new unknowns and raising important new questions. Avery himself had, with quintessential modesty, touched upon those important new questions in his letter to his brother:

If we are right, and of course that is not yet proven, then it means that nucleic acids are not merely structurally important but functionally active substances in determining the biochemical activities and specific characteristics of cells – and that by means of a known chemical substance it is possible to induce predictable and hereditary changes in cells. This is something that has long been the dream of geneticists . . . Sounds like a virus – may be a gene. But with mechanisms I am not now concerned – one step at a time – and the first is, what is the chemical nature of the transforming principle? Someone else can work out the rest . . .

three

The Story in the Picture

You look at science (or at least talk of it) as some sort of demoralising invention of man, something apart from real life, and which must be cautiously guarded and kept separate from everyday existence. But science and everyday life cannot and should not be separated.

ROSALIND FRANKLIN

The discovery of the 'transforming substance' by Avery, MacLeod and McCarty, confirmed by Hershey and Chase's elegant experiment with the bacteriophage, proved that DNA was the molecule of heredity. But both groups were working with microbes, bacteria and viruses, which were known to be much simpler in their hereditary nature than, say, animals and plants. This left huge unknowns that needed to be explored. Was DNA the key to the heredity of all of life, or was it just relevant to bacteria and viruses? By the early 1950s, work in many different laboratories had confirmed that DNA was a major ingredient in the nuclei of animals and plants. This supported the idea that DNA was the coding molecule of life. But if so, how did it really work? How, for example, did a single chemical molecule code for the complex heredity of a living organism?

Biologists, doctors, molecular biochemists and geneticists were now asking themselves the same, or similar, questions. Critical to any such understanding was the precise molecular structure of DNA. If, for example, we were to regard the role of DNA as akin to a stored genetic memory, how did that molecular structure enable the quality of such a phenomenally complex memory? How was that genetic memory transferred from parents to offspring? How did the same stored memory explain embryological development, where a single cell arising from the genomic union of a paternal sperm and maternal ovum gives rise to the developing human embryo and future adult human being?

There was another profoundly important question.

Darwinian evolution lay at the heart of biology. To put it simply, Darwin's idea of natural selection implied that nature selected from a range of variations in the heredity of different individuals within a species. The way in which it worked was exceedingly simple, if brutal. Those individuals, and by inference their variant heredities, who carried a small advantage for survival and thus a better chance of giving rise to offspring, would therefore be more likely to contribute to the species gene pool. In reality natural selection worked more through a process of attrition. Those less advantaged individuals who did not carry the advantage for survival, were more likely to perish in the struggle for existence, and thus they were less likely to contribute to the species gene pool.

This is what Darwinian evolutionary biologists refer to as 'relative fitness'. It is the measure of the individual's contribution to the species gene pool. Certainly it has nothing to do with racist notions of superiority and inferiority attached to 'survival of the fittest' – a term introduced not by Darwin but by the social philosopher Herbert Spencer. But if we take a pause and think about it, such variant heredity, essential for natural selection to work, must also come about through mechanisms involving this wonder molecule, DNA, which must lie not only at the heart of heredity but also at the absolute dead centre of evolution. All of these questions needed to be answered by the scientists now struggling to understand the structure and, assuming structure was function, the properties of this remarkable chemical, DNA.

In fact the first step towards answering these questions had already been taken back in 1943, in what might appear unlikely circumstances. It was taken not by a biochemist, biologist or geneticist, but by an Austrian physicist. The spark was lit when, at 4.30pm on Friday 5 February, Erwin Schrödinger stepped up to the podium in Dublin to deliver a lecture that is now seen as a landmark moment in the history of biology. Schrödinger had been awarded the Nobel Prize in 1933 for work in quantum physics that expanded our understanding of wave mechanics – but I won't confuse myself or my readers by entering further into the physics. The simple facts were that Schrödinger had exiled himself from his native Austria in protest at human rights abuses and had been given sanctuary in neutral Ireland by its President, Eamon de Valera. In Dublin Schrödinger had helped found the Institute for Advanced Studies. As part of his duties in support of the Institute, he had agreed to give a series of three lectures in which he developed a central theme: 'What Is Life?'

Such was Schrödinger's fame that the lecture theatre, which had a seating capacity for 400, could not accommodate all who wished to attend the lectures – this despite the fact that they had been warned in advance that the subject matter was a difficult one and that the lecture was not going to be pitched at an easy or popular level, even though Schrödinger had promised to eschew mathematics. De Valera himself was present in the audience, as were his cabinet ministers and a reporter for *Time* magazine. One wonders what these politicians and journalists made of Schrödinger's focus on 'how the events in space and time which take place within the spatial boundary of a living organism can be accounted for by physics and chemistry'.

Schrödinger subsequently extrapolated the three lectures into a book of less than a hundred pages with the same title: *What Is Life?* This was published the following year. In what is now a very famous book, Schrödinger popularised a quantum mechanics interpretation of the gene that had been proposed earlier by another distinguished physicist, the previously mentioned Max Delbrück.

In the opening pages of the first chapter, Schrödinger posed the question: 'How can the events which take place within a living organism be accounted for by physics and chemistry?' Admitting that at the time of writing the prevailing knowledge within the disciplines of physics and chemistry was inadequate to explain this, he nevertheless hazarded the opinion that 'the most essential part of a living cell – the chromosome fibre – may suitably be called *an aperiodic crystal*'. The italicisation is Schrödinger's to emphasise, as he further explained, that the physics up to this time had only concerned itself with periodic crystals, the kind of repetitive atomic structures seen, for example, in very obvious crystalline compounds such as gemstones.

What did he mean by an 'aperiodic crystal'?

He explained this with a metaphor. If we examined the images within the pattern of a wallpaper, we could see how the pattern was repeated, over and over. This was the equivalent of a periodic crystal. But if we examined the complex elaboration of a Raphael tapestry, we saw a pattern of images that did not repeat themselves, yet the pattern was coherent and meaningful.

Schrödinger intuited further.

It was the chromosomes, or more likely an axial fibre much finer than what was visible under the microscope, that contained what he termed 'some kind of code-script' that determined the blueprint of the individual's development from fertilised egg to birth – and further determined the functioning of what we would now term the genome throughout the lifetime of the individual.

That intuition would provide the drive for a naïve but highly inquisitive young American, called James Dewey Watson, to join forces with a slightly older but equally inquisitive Englishman, Francis Crick, and form what is now seen as one of the most famous partnerships in scientific history. Both men would take their inspiration from Schrödinger to search for the aperiodic crystal that coded for DNA.

*

Watson was an exceptionally bright child who lived at home with his family in Chicago while attending the local university. He enrolled when aged just 15 and he graduated, aged 19, in 1947 with a bachelor's degree that included a year studying zoology. His teacher of embryology would remember him as a student who showed little interest in lectures and made no notes whatsoever, so it was all the more puzzling when he graduated top of his class. Watson would subsequently admit to a habitual laziness. Though vaguely interested in birds, he had deliberately avoided any courses that involved chemistry or physics of 'even medium difficulty'. This self-indulgent student left Chicago with only a rudimentary knowledge of genetics or biochemistry. As part of his education he had attended lectures by the geneticist Sewall Wright, who had devised a mathematical system of studying population genetics. Wright's course included a discussion of Avery's work, but Watson would subsequently confess that he took little notice. He would also confess that the inspiration for his subsequent interest in the 'mystery of the gene' was Schrödinger's book, *What Is Life?*

Inspired by this book, Watson landed a research fellowship at Indiana University, at Bloomington. He was delighted by the move because Nobel Laureate Hermann Joseph Muller was the local Professor of Zoology. As early as 1921 Muller had observed that the genes of the fruit fly underwent mutations – as did the genes of the bacteriophages – the viruses that had inspired Hershey and Chase. Watson was intrigued by the fact that phage viruses could be manipulated in test tubes. Their reproductive cycles were extremely brief – an important consideration for an impatient young scientist. There were simple test systems that could be employed to follow their life cycles, and numbers, in a way that would open up new angles from which to attack the gene problem. All you had to do was carefully design an experiment aimed at probing some particular aspect of the gene problem and the whole shebang could be completed in a matter of days. This intimate, if brutal, interplay between phage viruses and their host bacteria allowed scientists to figure the complex chemistry of genes, genetics and chromosomes.

Curiously it would not be Muller but another phage researcher, Salvador Luria, who would now give shape and direction to the young scientist's growing infatuation with the gene.

The Italian-born Luria was another European scientist – a microbiologist, like Avery – who found refuge in America from the European war zone. By now he had entered into a working collaboration with Max Delbrück, who was Professor of Biology at the California Institute of Technology. In 1943 Luria and Delbrück designed and conducted an experiment that demonstrated that genetic inheritance in bacteria followed precise evolutionary principles. This experiment became one of the foundation stones of modern Darwinism. That same year Delbrück befriended another microbiologist called Alfred Hershey, who would subsequently write the key DNA paper with Martha Chase. In a letter to Luria, Delbrück summarised Hershey as follows: 'Drinks whiskey but not tea. Likes living in a sailboat . . . Likes independence.' The three scientists joined forces to become the nucleus of a cooperating and mutually supportive network of scientists that would become known as the 'phage group'. Delbrück would subsequently explain that they would be a group only in the sense that they communicated freely on a regular basis, and that they told one another what they were thinking and doing. In this way a loose creative movement grew around the two European expatriate scientists, all working towards the common ambition of figuring out how genes worked.

Luria, Delbrück and Hershey now posed some interesting questions. How does the phage virus actually get into the bacterium? How, once inside, does it multiply? Does it multiply like a bacterium, growing and budding off daughter viruses? Or does it multiply by an entirely different mechanism? Is this multiplication some complex physical or chemical process that could be understood in terms of known physical and chemical principles? Through making use of the phage reproductive system, they hoped to solve the mystery of the gene. To begin with it all seemed simple in principle, but as experiment followed experiment and year followed year, they found themselves no closer to the answer.

Up to 1940 or so, people like Delbrück and Luria had assumed that viruses were simple. They had little to go on since the majority of viruses were so minuscule they could not be seen with any clarity through the ordinary light microscope. They would even talk about them as if they were akin to protein molecules. Luria would come to define phage viruses, in a misleading oversimplification, as extensions of the bacterial genome. But with the invention of the electron microscope, by the German company Siemens, even the smallest viruses, including bacteriophages, would soon become visible for the first time. And when they did become visible, they proved to be more complex than the two scientists had initially conceived.

Many phages had a head that was cylindrical in shape, with a narrow sheath below it, as tall as the head, and a base plate with six spikes with fibres attached. Now that they could visualise phages in the process of infecting their host bacteria, something struck Delbrück and Luria as exceedingly odd. The viruses didn't actually pass through the bacterial cell wall. What they appeared to do was to squat down against the wall and inject their hereditary material into the cell. In 1951 a phage researcher called Roger Herriott would write to Hershey, 'I've been thinking that the virus may behave like a little hypodermic needle full of transforming principles.' This became the background to Hershey and Chase's experiment in which they confirmed that that was precisely what happened. The virus behaved exactly like a hypodermic syringe; the tail and its elongated fibrils would attach to the bacterial wall and the phage would then inject its viral DNA in through the bacterial wall to take over the bacterium's own genetic machinery, the viral genome compelling the bacterial genome to construct what was necessary for the generation of daughter viruses. In effect, the infected bacterium became a factory for the production of daughter viruses.

It would be this discovery, together with many associated extrapolations to microbiology and genetics, that would lead to all three scientists – Delbrück, Luria and Hershey – sharing the Nobel Prize in 1969.

Meanwhile, back in 1947, it was the dynamic energy and infectious charm of Luria, and the innovative genius of Delbrück, that proved most influential to the youthful Watson after his arrival into Indiana University. Still fascinated by the mystery of the gene, it was his hope that the mystery might be solved without his bothering to learn any of the complex physics or chemistry.

It is instructive to discover, from conversations between Luria and Watson, that there was no ignorance at Bloomington about

Avery's discovery of DNA. Luria had visited Avery in 1943, prior to the publication of the key paper, when he had the opportunity of discussing Avery's findings in detail. He would recall Avery to Watson as an utterly non-pompous scientist, precise in his language, with a tendency as he spoke to close his eyes and rub his bald head – 'every bit of a chemist, even though he was an MD'. Watson would take his cue from Luria, writing, in *The Double Helix*, how Avery had shown that hereditary traits could be transmitted from one bacterial cell to another by purified DNA molecules. Given the fact that DNA was known to occur in the chromosomes of every type of living cell, 'Avery's experiments strongly suggested that . . . all genes were composed of DNA.'

In the autumn of 1947, Watson, still just 19, took Luria's course in bacteriology and Muller's in X-ray-induced gene mutation. Faced with the choice of entering into research with Muller on Drosophila or with Luria on microbes, he plumped for Luria, despite the fact that the Italian scientist had a reputation among the graduate students for having a short fuse with dimwits. Watson would subsequently adopt his patron's example. Delbrück was a heroic figure to Watson because he had inspired Schrödinger's ideas in the inspirational book. Watson was delighted when Luria introduced him to Delbrück when the eminent German physicist paid a visit to Bloomington.

Luria set Watson a PhD dissertation on the pathological effects on phage of exposure to X-rays. The work proved so mundane that Watson would barely mention it in his biography. But his obsession with the gene was undimmed. By the summer of 1949, his thesis nearing completion, he had the itch to travel to Europe. Luria arranged a Merck Fellowship from the National Research Council – three thousand dollars for the first year, potentially renewable. In May the following year, with his PhD under his belt, he sailed for Denmark, where he had been assigned to study nucleotides with a biochemist named Herman Kalckar. Kalckar was a gifted scientist but his interest was neither the gene nor the bacteriophage. A disenchanted Watson switched his attentions to another Dane, and a member of the phage group, Ole Maaløe, who was working on the transfer of radioactively-tagged DNA from phages to their viral offspring.

Out of the blue, Kalckar accepted a short-term project in the Zoological Station in Naples. He suggested that Watson might tag along. Though he had little interest in marine biology, Watson was delighted to acquiesce. He hoped to warm himself in the Italian sun. But he was disappointed to find Naples chilly, with no heater in his room on the sixth floor of a nineteenth-century house. 'Most of my time I spent walking the streets or reading journal articles . . . I daydreamed about discovering the secret of the gene, but not once did I have the faintest trace of a respectable idea.'

Here, by happenstance, he attended a lecture in the Zoological Station given by an English scientist named Maurice Wilkins. The lecture could hardly have excited him in prospect, knowing that most of it would be about the biochemistry of proteins. 'Why should I get excited learning boring chemical facts as long as the chemists never provided anything incisive about the nucleic acids?'

But he took the risk and attended anyway.

Tall, bespectacled, asthenic and somewhat diffident in manner, you might have expected Wilkins' presentation to bore the restless and impatient Watson. But it did not. To begin with, it was delivered in a language that Watson readily understood. And for all of his diffident manner, Wilkins kept to the point. Then suddenly, close to the end of the lecture, a projected slide jarred Watson to full attention. On the screen was a photograph that showed something Wilkins called an X-ray diffraction pattern of DNA that had been taken in the King's College laboratory in London. Watson would subsequently admit that he knew nothing about X-ray crystallography. He hadn't understood a word of what he had read about it in the scientific journals and he thought that much of what the 'wild crystallographers' were claiming was very likely baloney.

But now here was Wilkins mentioning in passing that this was the clearest picture of DNA that he and his colleagues had yet obtained from their X-ray studies. In the same audience was the Leeds-based English physicist, William Astbury, who had pioneered X-ray diffraction studies of biological molecules, and who had produced the first X-ray pictures of DNA. Astbury would subsequently confirm that no one had ever shown such a sharp, discrete set of reflections from the DNA molecule as Wilkins then projected onto the screen. 'There was nothing like it in the literature.' In explaining the picture, Wilkins suggested that DNA might be thought of as a crystalline substance.

Watson was electrified to hear Schrödinger's prophecy confirmed. He sat in a daze of wonderment as Wilkins went on to explain that if and when we understood the structure of DNA, then we might be in a better position to understand how genes worked. Watson was now asking himself some pertinent questions. Who was this interesting English scientist, Wilkins? And how could he get to join his team at King's College in London?

*

Maurice Hugh Frederick Wilkins was not, in fact, English, as Watson initially surmised. He was born in Pongaroa, New Zealand, where his father, Edgar Henry, was a practising doctor. The family were Anglo-Irish in origins, hailing from Dublin, where Maurice's paternal grandfather had been headmaster of the high school and his maternal grandfather chief of police. On leaving New Zealand the family first returned to Ireland, then headed for London, where Dr Wilkins was later to do his pioneering work in public health.

Maurice had had a natural scientific curiosity even as a boy, and it was this curiosity that led to his studying physics as part of his BA at Cambridge University, after which he worked for his PhD under John Turton Randall (later knighted), a physicist who played a leading role in the development of radar during the war.

As a postgraduate, Wilkins moved to the University of Birmingham, following the posting of his Cambridge tutor, Randall, where the two scientists continued their collaboration on radar. But then, out of the blue, Wilkins found himself dispatched to the United States to work on the Manhattan Project. His purpose was to figure out how to purify suitable isotopes of uranium from impure sources, to make them suitable for the atomic bomb. In February 1944 Wilkins crossed the dangerous waters of the Atlantic on the *Queen Elizabeth*, heading for the University of Berkeley, California. Here he made a modest contribution to the development of the atomic bomb. However, the subsequent destruction of Hiroshima and Nagasaki by the very weapons that he had worked on left Wilkins somewhat unsettled in conscience.

After the war Wilkins returned to England, where he ended up as assistant director of the new Biophysics Unit at King's College London, funded by the Medical Research Council, and where his former boss, Randall, was now the Wheatstone Professor of Physics. The new departmental remit was to apply the experimental methods of physics to important biological problems. This would result in Wilkins developing a relationship with Watson and Crick and joining the search for the molecular code of DNA. It would also involve him in a somewhat infamous strained working relationship with the X-ray crystallographer Rosalind Franklin.

Given this developing history, we might pause a moment or two to consider Wilkins' personality, and its relevance to the coming storm. From what one can gather from his belatedly published biography, and the memory of those who knew him and worked with him, Wilkins was a quiet, highly moral man, somewhat Quaker-like in social attitudes. As a boy he enjoyed a close emotional relationship with his elder sister, Eithne, who taught him to dance. But this intimacy was torn apart when Eithne developed a bacterial infection that turned into a septicaemia, the blood-borne infection provoking septic arthritis in multiple joints. This would have been a shockingly painful and disabling condition, which, prior to antibiotics, might have proved fatal. She spent months in a hospital bed, with her limbs dangling from hoists, her joints lanced open to drain the pus. The unfortunate Eithne survived but the intimacy with her younger brother ended. The trauma of this experience may well have affected his self-confidence, particularly in his relationships with women.

While an undergraduate at Cambridge, he fell in love with a woman called Margaret Ramsey, but he 'was incapable of making a suitable advance to her'. After he told her of his love, there was a short silence after which she walked from the room. During his stay in Berkeley, Wilkins was attracted to an artist named Ruth, who had shared lodgings with him. She conceived a child and they subsequently married, but when, as the war was ending, he informed Ruth that he intended to return to the UK, she refused to accompany him. 'Ruth told me one day that she had made an appointment for me with a lawyer and when I arrived at his office I was shocked to hear that Ruth wanted to end our marriage.' Shortly after the divorce, Ruth gave birth to a son. Wilkins went to see her, and their baby, in the hospital ward, before returning to the UK alone.

Wilkins would admit to difficulty overcoming an innate shyness, and he would require periodic psychotherapy in his time working at King's, but he subsequently found a wife, Patricia, who appreciated the sensitive soul behind the diffident exterior, and he enjoyed a happy marriage and the joys of rearing a family of four children. There was also a fruitful outcome of his unsettled conscience following his work on the Manhattan Project. Before leaving Berkeley, one of his working colleagues came to his rescue . . . 'Seeing I wanted to find some new direction, he lent me a new book with the rather ambitious title, *What Is Life?*'

49

four

A Couple of Misfits

Francis likes to talk ... He doesn't stop unless he gets tired or he thinks the idea's no good. And since we hoped to solve the structure by talking our way through it, Francis was the ideal person to do it.

JAMES WATSON

It is somewhat ironic that Maurice Wilkins only arrived in Naples by happenstance, since he was substituting for Randall, who had agreed to present the talk but had been unable to attend. It seems unlikely, had Randall himself presented the lecture, that he would have included the DNA slide, or that he would have spoken of what it portrayed with such clear reference to Schrödinger's book. This lecture, which so excited Watson, was on the physico-chemical structure of big biological molecules, mostly proteins, made up of thousands of atoms. The key photograph had been taken by Wilkins, working together with a graduate student called Raymond Gosling while using a technique called X-ray diffraction. One of the things this technique was particularly good at was finding the sort of repetitive molecular themes you found in crystals, hence the other term for it: X-ray crystallography.

'Suddenly,' as Watson would later recall, 'I was excited about chemistry.'

Up to this moment Watson had had no idea that genes could crystallise. To crystallise, substances must have a regular atomic structure – a lattice-like structure of atoms at the ultramicroscopic level. The youthful Watson appears to have been a wonderfully free spirit journeying from one interesting encounter to another. Impulsive, impatient, egregiously direct, yet all the while on the hunt for new adventure.

'Immediately I began to wonder whether it would be possible for me to join Wilkins in working on DNA.' But Watson never got to work with Wilkins. Instead, happenstance headed him in the direction of another X-ray crystallographer called Max Perutz, who was working at the Cavendish Laboratory at Cambridge University.

The Cavendish Laboratory is a world-famous department of physics. First established in the late nineteenth century to celebrate the work of British chemist and physicist Henry Cavendish, one of its founders and the first Cavendish Professor of Physics was James Clerk Maxwell, famous for his development of electromagnetic theory. The fifth Cavendish Professor and the director of the laboratory at the time of Watson's arrival was William Lawrence Bragg, who was the successor, as director, to Lord Ernest Rutherford, another Nobel Prize-winner and the first physicist to split the atom. Bragg was an Australian-born physicist who, jointly with his father, had been awarded the Nobel Prize in Physics in 1915 for establishing the use of X-rays in analysing the physico-chemical structures of crystals. X-ray beams are bent when they pass through the orderly atomic lattice of crystals. What is projected onto the photographic plate is not the picture of the atoms within the structure but the refracted pathways of the X-rays after they have collided with the atoms. This is called 'diffraction' and is similar to how light is bent when it passes through water. In a structure with haphazard positioning of atoms in space, the X-rays will be scattered randomly and form no pattern. But in a structure that contains atoms in a repetitive atomic lattice – such as a crystal – the X-rays are deflected in a recognisable pattern of blobs on the X-ray plate. From this diffraction pattern, the atomic structure of the structure can be deduced.

The two Braggs – father and son working as a team at the University of Leeds – had constructed the first X-ray spectrometer, allowing scientists to study the atomic structure of crystals. At the age of 22, Bragg Junior, now a Fellow of Trinity at Cambridge, had produced a mathematical system, Bragg's Law, that enabled physicists to calculate the positions of the atoms within a crystal from the X-ray diffraction pictures. At the time of Watson's arrival into the laboratory, Bragg's main focus of study was the structure of proteins. It was this potential for the X-ray diffraction of proteins that had attracted Max Perutz to the Cavendish Laboratory.

Born in Vienna of Jewish parentage, Perutz was another enforced exile who had settled in England and become a research student at the Cavendish Laboratory. He completed his PhD under Bragg and subsequently devoted most of his professional life to the analysis of the macromolecule of haemoglobin, the pigment that colours the red cells in our blood, enabling them to carry oxygen around the body. Also working at the Cavendish was an unusual young scientist, Francis Crick. The English-born scientist had graduated with a BSc in physics from University College London aged 21, but thanks to war duty and a profound antipathy to his PhD project (he was supposed to be working on the viscosity of water at high temperatures) he, like Watson, found an alternative source of inspiration in Schrödinger's book. In Crick's own words, 'It suggested that biological problems could be thought about in physical terms.'

But what terms?

At the time Crick wasn't as convinced by Avery's discovery as Watson was. Like Schrödinger himself, Crick was more inclined to the protein hypothesis. But he was every bit as impressed with Schrödinger's 'code-script' idea as Watson. What then could he possibly make of Schrödinger's conception of an aperiodic crystal?

Simple crystals such as sodium chloride, the basis of common salt, would be incapable of storing the vast memory needed for genetic information because their ions are arranged in a repetitive or 'periodic' pattern. What Schrödinger was proposing was that the 'blueprint' of life would be found in a compound whose structure had something of the regularity of a crystal, but must also embody a long irregular sequence, a chemical structure that was capable of storing information in the form of a genetic code. Proteins had been the obvious candidate for the aperiodic crystal, with the varying amino acid sequence providing the code. But now that Avery's iconoclastic discovery had been confirmed by Hershey and Chase, the spotlight fell on DNA as the molecular basis of the gene. Suddenly new vistas of understanding the very basics of biology, and medicine, appeared to be beckoning.

It was through a mixture of luck and the gut reaction of Perutz that the dilettantish Crick was taken into the fold of the Cavendish. In Perutz's recollection, Crick arrived in 1949 with no reputation whatsoever in science. 'He just came and we talked together and John Kendrew and I liked him.' And so the likeable Crick ended up, in such an idiosyncratic process of selection, working on the physical aspects of biology – what today we call molecular biology – under the guidance of Bragg, Perutz and Kendrew, at the Cambridge laboratory.

In 1934, John Desmond Bernal, an Irish-born scientist with Jewish ancestry and a student of Bragg Senior, had shown for the first time that even complex organic chemical molecules, such as proteins, could be studied using X-ray diffraction methods. Bernal was a Cambridge graduate in mathematics and science, who was appointed as lecturer to Bragg at the Cavendish in 1927, becoming assistant director in 1934. Together with Dorothy Hodgkin, Bernal pioneered the use of X-ray crystallography in the study of organic chemicals – the chemicals involved in biological structures – including liquid water, vitamin B1, the tobacco mosaic virus and the digestive enzyme, pepsin. This was the first protein to be examined at the Cavendish in this way. When, in 1936, Max Perutz arrived as a student from Vienna, he extended Bernal's work to the X-ray study of haemoglobin.

By the time Crick joined the laboratory, Sir William Bragg had been replaced by Sir Lawrence Bragg, and John Kendrew and Max Perutz had taken Bernal's findings further to become bogged down in a 'disastrous paper' on the chain structures of proteins. And now we discover something distinctly unusual about Francis Crick, something that Perutz may have intuited at their meeting. He had an avid curiosity about science, reading very widely, and he was equipped with a mind capable of amassing a formidable knowledge base across different disciplines. One of the first things he did after his arrival into the Cavendish was to acquaint himself with everything his bosses had achieved. Junior as he was, Crick now took it upon himself to undertake a long, critical look at their work. This he then proceeded to criticise from basic principles. At the end of his first year in the department, Crick presented his criticisms in the form of an ad hoc seminar, borrowing his title from Keats as 'What Mad Pursuit'. He began with a twenty-minute summary of the deficiencies in the departmental methods before pointing out what he saw as the 'hopeless inadequacy' of their investigation of the structure of the haemoglobin molecule. The X-ray analysis of haemoglobin was of course Perutz's main objective. Bragg was infuriated by the cocky behaviour of this upstart junior colleague, but Perutz would subsequently admit that Crick was right and proteins were far more complicated in their structures than they had initially assumed. Restless and ever-inquisitive, Crick proved to be an uneasy, sometimes downright embarrassing import into the scientific pool of the laboratory. And while Bragg and Perutz saw proteins as the great unsolved puzzle, Crick was more interested in the mystery of the gene.

As 1949 elided into 1950, Crick would subsequently confess that he still did not realise that the genetic material was DNA. But he knew that genes had been plotted out in linear arrays along the chromosomes by people like Barbara McClintock, and that proteins, which had to be the expression of the genes, were also being plotted out as linear arrays, however lengthy and complicated. There had to be some logical way in which one translated into the other. By 1951, two years after his arrival into the Cavendish Laboratory, Crick perceived that these were two different, if necessarily related, puzzles – the mystery of how genes appeared able to copy themselves, and the mystery of how the linear structures of genes translated into the linear structures of proteins.

The wide-reading, voraciously inquisitive Crick needed what Judson termed a catalyst. This arrived in the form of the gangly, equally inquisitive Watson that same year, 1951. From their first meeting, it would appear that here was one of those rare working conjunctions of two odd-ball personalities that, when they come together, make an extraordinary creative whole that is more than the sum of the individual ingenuities. And yet it very nearly didn't happen.

*

We should recall that Watson was extremely junior within the department. A recent PhD graduate, he had arrived into Kalckar's laboratory on a Merck Fellowship funded by the US National Research Council. The terms and conditions were laid down and signed for back home, but now here he was abandoning those carefully laid intentions to gallivant from the work in Denmark to follow some giddy new inspiration in England, a place he had never visited in his life and where he knew absolutely nobody. Impulsive and single-minded, Watson would subsequently confess that his head was filled with curiosity about that single DNA photograph. He had tried to engage with Wilkins in Naples after the lecture, at a bus stop during an excursion to the Greek temples at Paestum. He had even tried to take advantage of a visit from his sister, Elizabeth, who had arrived to join him as a tourist from the States. Now here were Maurice Wilkins and Watson's sister, Elizabeth, finding a common table to take lunch together. Watson sensed an opportunity and barged in, with the intention of ingratiating himself with Wilkins. But the self-effacing Wilkins excused himself, to allow brother and sister the privacy of the table.

His plans foiled, Watson refused to let go of this exciting new avenue of interest. 'I proceeded to forget Maurice, but not his DNA photograph.'

He stopped over in Geneva for a few days to talk to a Swiss phage researcher, Jean Weigle, who provoked yet more excitement by informing Watson that the eminent American chemist, Linus Pauling, had partly solved the mystery of protein structure. Weigle had attended a lecture by Pauling, who like Bragg in Cambridge had been working with X-ray analysis of protein molecules. Pauling had just made the announcement that the protein model followed a uniquely beautiful three-dimensional form – he had called it an 'alpha-helix'. By the time Watson arrived back in Copenhagen, Pauling had published his discovery in a scientific paper. Watson read it. Then he re-read it. He was confounded by his lack of understanding of X-ray crystallography. The terminology, in physics and chemistry, was so far beyond him that he could only grasp the most general impression of its content. His reaction was so childishly naïve as to be touching: in his head he devised the opening lines of his own imagined paper in which he would write about his discovery of DNA, if and whenever he discovered something of similar portent.

But what to do to get on board the DNA gravy train?

He needed to learn more about X-ray diffraction studies. Ruling out Caltech, where Pauling would react with disdain to some 'mathematically deficient biologist', and now ruling out London, where Wilkins would be equally uninterested, Watson wondered about Cambridge University, where he knew that somebody called Max Perutz was following the same X-ray lines of investigation of the blood protein molecule, haemoglobin.

'I thus wrote to Luria about my newly found passion . . .'

The world of science was smaller in 1951 than it is today. Even so, it would appear a hopelessly optimistic ambition for this impulsive young graduate to merely ask his mentor to fix his arrival into a leading laboratory in England to engage in a line of research that he knew absolutely nothing about.

The amazing outcome was that Luria was able to do so. By happenstance, he met Perutz's co-worker, John Kendrew, at a small meeting at Ann Arbor, in Michigan, where, by a second and equal happenstance, there was a meeting of minds – both scientific and social. And by a third happenstance, Kendrew was looking for a junior to help him study the structure of the muscle-based protein myoglobin, which contained iron at its core and held on to oxygen, just like the haemoglobin in the blood.

Twice in his short career the young American scientist had leapt into the unknown and landed on his feet. First it had been through Luria's patronage in Bloomington, and by extension also Delbrück's, two of the co-founders of the phage group; and now the gift of happenstance extended further, again through Luria's patronage, to Kendrew, and by proxy to the Cambridge laboratory and Max Perutz. Watson's arrival into the laboratory would bring him under the ultimate tutelage of Sir Lawrence Bragg, a founder of X-ray crystallography. It would connect him directly to his future partner in DNA research, Francis Crick, and further afield – through the connection between the Cambridge laboratory and the X-ray laboratory at King's College London – with Maurice Wilkins and a young female scientist, Rosalind Franklin, who were working on the X-ray crystallography of DNA.

five

The Secret of Life

I think there was a general impression in the scientific community at that time that [Crick and Watson] were like butterflies flicking around with lots of brilliance but not much solidity. Obviously, in retrospect, this was a ghastly misjudgement.

MAURICE WILKINS

In the opening pages of his brief, witty and brutally candid autobiography, James Watson recounts a chance meeting in 1955 with a scientific colleague, Willy Seeds, at the bottom of a Swiss glacier. It was two years after the publication of the discovery of DNA. Watson and Seeds were acquainted, Seeds having worked with Maurice Wilkins in probing the optical properties of DNA fibres. Where Watson had anticipated the courtesy of a chat, Seeds merely remarked, 'How's Honest Jim?', before striding away. The sarcasm must have bitten deep for Watson to not merely remember it distinctly, but even to consider the term 'Honest Jim' as the initial title of his life story, before being persuaded to adopt the more descriptive alternative, 'The Double Helix'. It was as if the former colleague was questioning Watson's right to be recognised as the co-discoverer of the secret of life.

He had been taken aback, reflecting on meetings with the same colleague in London a few years earlier, at a time when, in Watson's words, 'DNA was still a mystery, up for grabs . . . As one of the winners, I knew the tale was not simple, and certainly not as the newspapers reported.' It was a more curious story, one in which his fellow-discoverer, Francis Crick, would freely admit that neither he nor Watson was even supposed to working on DNA at the time. Equally curious was the fact that up to the day of the discovery, neither Watson nor Crick had contributed anything much to the many different scientific threads and themes that, when finally put together, like the pieces of a remarkable three-dimensional jigsaw puzzle, laid the molecular nature of DNA bare for the first time in history.

Watson's welcome into the Cambridge laboratory was quintessentially English in its lack of formality. He arrived in Perutz's office straight from the railway station. Perutz put him at his ease about his prevailing ignorance of X-ray diffraction. Both Perutz and Kendrew had come to the science from graduation in chemistry. All Watson needed to do was to read a text or two to become acquainted with the basics. The following day Watson was introduced to the white-moustached Sir Lawrence, to be given formal permission to work under his direction. Watson then returned to Copenhagen to collect his few clothes and tell Herman Kalckar about his good luck. He also wrote to the Fellowship Office in Washington, informing them of his change of plans. Ten days after he had returned to Cambridge he received a bombshell in the post: he was instructed, by a new director, to forgo his plans. The Fellowship had decided he was unqualified to do crystallography work. He should transfer to a laboratory working on physiology of the cell in Stockholm. Watson appealed once more to Luria.

As far as Watson was concerned it was out of the question to follow these new instructions. If the worst came to the worst, he would survive for at least a year on the $1,000 still left to him from the previous year's stipend. Kendrew helped him out when his landlady chucked him out of his digs. It was just another indignity when he ended up occupying a tiny room at Kendrew's home, which was unbelievably damp and heated only by an aged electric heater. Though it looked like an open invitation to tuberculosis, living with friends was preferable to the sort of digs he might be able to afford in his impecunious state. And there was a comfort to be had:

'I had discovered the fun of talking to Francis Crick.'

And talk they did.

In Crick's own memory: 'Jim and I hit it off immediately, partly because our interests were astonishingly similar and partly, I suspect, because a certain youthful arrogance, a ruthlessness, and an impatience with sloppy thinking came naturally to us both.' That conversation, lasting for two or three hours just about every day for two years, would unravel the most important mystery ever in the history of biology – the molecular basis of heredity.

We need to grasp a few fundamentals to understand how this happened. Firstly, we have two young and ambitious men – in Watson's case aged just 23, in Crick's, aged 35 – who were both exceptionally intelligent and surrounded by the ambience of high scientific endeavour and achievement. We need to grasp that Watson's interest, intense and obsessive, was the structure of DNA in its potential to explain the mystery of the workings of the gene, and thus the storing of heredity. We also need to grasp the slight, but important, difference with Crick's interest, which was not DNA, or even the gene in itself, but the potential of DNA to explain how Schrödinger's mysterious molecular codes – his aperiodic crystals – had the potential not only for coding heredity but for translating from one code to another, from the gene to the second aperiodic crystal that must determine the structure of proteins.

Crick would subsequently recall Watson's arrival in early October 1951. Odile, his French second wife, and he were living in a tiny ramshackle apartment with a green door that they had inherited from the Perutzes. Conveniently situated for the centre of Cambridge and only a few minutes' walk from the Cavendish Laboratory, it was all they could afford on Crick's research stipend. The 'Green Door', as it was thereafter called, consisted of an attic over a tobacconist's house, with 'two and a half rooms' and a small kitchen that was reached by climbing a steep staircase off the back of the tobacconist's house. The two rooms served as living room and bedroom for Crick and Odile, with the half room providing a bedroom for Crick's son, Michael – born to his first wife, Ruth Doreen – when Michael was home from boarding school. The wash-room and lavatory opened halfway up the stairs and the bath, covered with a hinged board, was a feature of the tiny kitchen.

One day, out of the blue, Perutz brought Watson to the flat. Crick was out. But he would recall Odile remarking that Max had come round with a young American who 'had no hair'. The newly arrived Watson was sporting a crew-cut – a hairstyle uncommon in England at the time. They met within a day or two . . . 'I remember the chats we had over those first two or three days in a broad sort of way.'

Both men were impecunious, but it hardly mattered since they were uninterested in money. What mattered was that the deeply personal, deeply intellectual, symbiosis had begun. Crick brought a rowdy enjoyment of problem solving, together with the hubris, born out of his background in physics, to believe that the big problem facing them – the mystery of the gene – was indeed solvable. Watson, who had little knowledge of physics or X-ray crystallography, brought a mine of knowledge about the way in which genes worked – the fruits of the bacteriophage researches of Luria and Delbrück. Perutz would subsequently confirm that the arrival of Watson, at that particular moment of time, was opportune for the workings of the Cavendish Lab, where his enthusiastic personality appeared to have galvanised Crick, and where his knowledge of the field of genetics added an exotic aspect to the structural physics and chemistry that otherwise prevailed. Moreover, different as their backgrounds were, Crick and Watson shared a deep, insatiable level of curiosity about the puzzle that lay at the very root of biology: they were determined, almost from their first meeting, that they would solve the mysterious nature of the gene.

The first creative step was to realise that the answer lay with DNA. To be more accurate, they realised that somehow chemical structure must parallel function: so the answer to the great conundrum lay in the three-dimensional chemical structure of DNA. But nobody really knew what shape or form this structure took. To the minds of Crick and Watson at that particular moment in time, it would have seemed nothing more than a ghost in the mist.

New discoveries in science will usually involve a lengthy period of laboratory labour, with knowledge growing by hard-won increments, often involving contributions from several, or a good deal more than several, different sources. In many ways the struggle to get to grips with the mysteries of heredity followed exactly such a course. But the mundane sweat of the laboratory aspects, the growth of knowledge by hard-won increments, would not fall to Watson and Crick. These would be left to others. The Crick–Watson symbiosis would be founded on a second, equally important ingredient of scientific advance, and one that has commonalities with the advances in the arts and humanities: this is the quintessentially human gift we call 'creativity'.

Within the hierarchy of the lab, Crick and Watson were the lowest contributing level. In Crick's words, 'I was just a research student and Jim was just a visitor.' They read very widely, imbibing the fruits of the hard work of others. They talked and talked, thinking out loud, probing one another's ideas and knowledge, often with Crick playing devil's advocate. In fact they gossiped and argued so much they were given a room to themselves – to avoid their interrupting the thoughts of their more senior colleagues – within the crowded structure of the old Cavendish Laboratory. The X-ray laboratory, with its heavy machinery and radiation dangers, was located in the basement. Jim and Francis would also share a cheap and cheerful lunch, of shepherd's pie or sausage and beans, at the local pub, the Eagle – a grubby establishment in a cobblestoned courtyard – where the creative debate would simply continue.

What little they knew about DNA was made even more uncertain by the fact that Crick believed that much of what was generally assumed to be the case with DNA and heredity was almost certainly wrong. It had been this attitude that had got him into trouble with Bragg. It meant that he didn't even trust the work of his seniors here in the lab. But the real reason behind Bragg's anger was his resentment of the fact that the chemist, Pauling, had discovered the alpha helix of protein. Meanwhile, Crick was convinced that the reason why the Cavendish had missed out on this was because they were assuming the accuracy of some earlier experimentation on the X-ray interpretation of the skin protein, keratin, which is the main ingredient of our human nails and a raptor's claws. The way in which Crick's mind worked can be gleaned from a remembered conversation:

'The point is [so-called] evidence can be unreliable, and therefore you should use as little of it as you can. We have three or four bits of data, we don't know which one is reliable . . . [What if] we discard that one . . . then we can look at the rest and see if we can make sense of that.'

*

Watson joined the Cavendish in the same year, 1951, in which Linus Pauling published his paper on the protein 'alpha helix'. This discovery so rattled Watson that all of the time he was working with Crick on the structure of DNA, he was looking over his shoulder in Pauling's direction.

He had good reason for seeing Pauling as the supreme rival in such an exploration; awarded the Nobel Prize in Chemistry in 1954, Pauling was already being hailed by scientific historians as one of the most influential chemists in history. His master work, though he contributed a great deal more, was to apply a quantum theory perspective to the chemical bonds that bind atoms within the structure of molecules, extending this basic science to the complex organic molecules that are the chemical building blocks of life.

The twentieth century has amazed us with its achievements in astronomy, in which scientists have plotted the stars and galaxies, and the forces, such as black holes, that govern the Universe. Equally important, though not so easily recognised as such by the ordinary man and woman, have been the achievements of the chemists and biochemists in exploring the micro-universe of atoms and molecules. Two forces in particular play a key role in the way that atoms bind to one another to make up life's particular molecules. One of these is called the covalent bond; the other is called the hydrogen bond. Pauling applied the science of quantum mechanics to the forces involved in these two very different chemical bonds.

We have no need to concern ourselves with the complex mathematics of the applied physics. We just need to grasp the basic mechanics. And where better to look than at the familiar molecule of water.

Everybody knows that the chemical formula for water is H_2O. This tells us that a molecule of water comprises one atom of oxygen and two atoms of hydrogen. But how do they link with one another to form the stable compound that we handle and consume every day of our lives? The molecule of water might be compared to a planet, oxygen, with two encircling moons of hydrogen. In such a situation, we can readily imagine how the force of gravity would hold the hydrogen moons to their orbits around the oxygen planet. In molecular terms, the forces holding the two hydrogen atoms to the oxygen atom are called 'covalent bonds'. At the ultramicroscopic level of atoms, the nucleus of each hydrogen atom contains a single positively charged proton while circling around the nucleus is a single negatively charged electron. Meanwhile, the oxygen atom has eight positively charged protons within its nucleus and eight balancing, negatively charged electrons in orbits around it. These electrons occupy two orbits – two electrons taking up an inner orbit and six taking up an outer orbit. In coming together to form a molecule of water, the two electrons in orbit around each of the two hydrogen nuclei have paired with two of the six electrons of the oxygen outer orbits. The paired electrons share their attraction to the protons of the two parent nuclei, so the paired electrons are now equally attracted to the oxygen nucleus and the hydrogen nuclei. This sharing of attraction creates a stable 'covalent' bond between the three atoms, just as gravity created stable orbits for the two moons rotating around our imaginary planet of oxygen.

Hydrogen bonds are something else.

Once again, we might take water as our example. But here we are looking at the chemical interactions between whole water molecules – the H_2Os reacting with one another. There are forces of attraction, albeit rather weaker and less stable than covalent bonds, between certain molecules that contain both hydrogen and heavier atoms such as nitrogen, oxygen or fluorine. Since water contains hydrogen and oxygen, these hydrogen bonds can form between molecules of water – it is this sticking together of water molecules that explains the difference between water vapour, or steam, liquid water and solid water, or ice. In ice most of the molecules are attached to one another by hydrogen bonds, to form something like a crystal; in liquid water varying amounts are attached to one another; and in steam, as a result of the addition of energy through heating, the hydrogen bonds linking water molecule to water molecule are broken down but the covalent bonds linking atoms to atoms remain intact.

We see that hydrogen bonds are weak, and thus unstable when heated, but covalent bonds are stable. These same two bonds, covalent and hydrogen bonds, are important ingredients in the structure of organic chemicals such as proteins. And they are also important in the structure of DNA.

Between 1927 and 1932, Pauling published some fifty scientific papers in which he conducted X-ray diffraction studies, coupled with quantum mechanical theoretical calculations, leading him to postulate five rules, known as Pauling's rules, that would help science to predict the nature of the bonds that held together atoms within molecules. At least three of these rules were based on Bragg's own work, the purloining of which provoked Bragg to fury. It was now inevitable that there would be ongoing scientific rivalry between the two scientists. Pauling's work into the nature of chemical bonding was so original, and pioneering, that he was awarded the Nobel Prize in Chemistry in 1954. Meanwhile, this new level of understanding enabled Pauling to visualise the precise shape and dimensions of molecules in three-dimensional space. Working at Caltech, Pauling applied this to the huge molecules of proteins, using the techniques of X-ray diffraction analysis pioneered by the Braggs. He showed, for example, that the haemo-globin molecule – the focus of Perutz's research – changed its physical structure when it gained or lost an oxygen atom. And Pauling continued to apply his rules to researching the molecular structure of proteins.

Pioneering X-ray pictures of fibrous proteins had been obtained some years before at the University of Leeds by William Thomas Astbury, the physicist who had attended Wilkins' talk in Naples, but it was assumptions based on these X-ray diffraction pictures that Crick was now questioning at the Cavendish Laboratory. For many years Pauling had tried to apply quantum mechanics calcu-lations to Astbury's X-ray pictures, but he found that things just didn't add up. It would take him and two collaborators, Robert Corey and Herman Branson, fourteen years before they made the necessary breakthrough.

All proteins have a primary structure that is made up of an amino acid code, with the letters made up of twenty different amino acids. The chemical bonds that join up the amino acids into the primary chain are called 'peptide bonds'. Pauling and his collaborators now realised that peptides bonded together in a flat two-dimensional plane – they called this 'a planar bond'. A problem with outdated equipment had caused Astbury to make a critical error in taking his X-ray pictures: the protein molecules became tilted away from their natural planes, skewing the mathematical extrapolations of their structure. Once they had corrected Astbury's error, Pauling and co discovered that as the chain of amino acids grew, to form the primary structure of proteins, it naturally followed the shape of a coiled spring, twisting to the right – the so-called 'alpha helix'. This was the discovery that had excited Watson on his return trip from Naples.

Back in Cambridge, Sir Lawrence Bragg was bitterly disappointed when Pauling's group beat his to the discovery of the primary structure of proteins. But there was a silver lining to the cloud: Perutz now used Pauling's breakthrough to reappraise his own work on the haemoglobin molecule, a reappraisal that would solve the structural puzzle of haemoglobin and garner his Nobel Prize in Chemistry in 1963. Pauling's discovery also alarmed Watson who, from his arrival at Cambridge, had assumed that they had a very knowledgeable and powerful rival in what was now a race to discover the three-dimensional structure of DNA.

*

But the problem, as Crick would point out in their day-to-day sharing of thoughts and incessant debate, was that they couldn't even assume that Pauling's data was right. In Crick's words, 'Data can be wrong. Data can be misleading.' So Crick and Watson attempted to construct their physical model with a sceptical eye on prevailing experimental data. To put it another way, they relied just as heavily on creative leaps of their own imagination as on existing experimental data.

Crick and Watson were now asking themselves if DNA, like proteins, had a helical structure, and Watson in particular was convinced that they should also take their cue from Pauling, who liked to construct three-dimensional models of the molecules he was attempting to envisage. To do so they would have to think, as Pauling did, about the atomic structures that made up the chemistry of DNA – to fit the molecules, with their component atoms, and the bonds between them, into a complex three-dimensional jigsaw. They knew that they were dealing with the four nucleotides – guanine, adenine, cytosine and thymine – together with the molecule of the sugar, called ribose, and the inorganic chemical, phosphate, all of which, when correctly fitted together, must somehow make up the mysterious three-dimensional jigsaw puzzle.

Two relevant questions now loomed. Firstly, if the structure was helical, what kind of helix was involved? And secondly, where did the phosphate molecule fit into the structure? Calcium phosphate is the mineral of bones, of shells, of rocks formed from the remains of living marine organisms – limestone. The presence of phosphate suggested some kind of strengthening of the DNA chain – a chemical scaffold – maybe a spine? But where did this spine lie in relation to the presumptive and as yet unknown spiral? And where, or how, did the sugar fit in? The code itself must surely lie with the nucleotides, acting perhaps as something like letters. Each was a key ingredient, but how on earth did the whole thing assemble in a way that made sense?

An important clue must come from the X-ray diffraction patterns. That meant they needed the help of Maurice Wilkins and Rosalind Franklin – 'Rosy', as Watson referred to her in his autobiography – who were conducting X-ray analyses of DNA fibres at the King's College London laboratory.

*

Rosalind Elsie Franklin was born in London to a prosperous Jewish family in 1920. From an early age she showed both a brilliantly incisive mind and the stubbornness necessary to make a distinguished mark for herself. She also showed an aggressively combative side to her personality that might prove a mixed blessing in overcoming the prevailing prejudices against Jews in society, as well as against women being in higher education and the scientific workplace. It didn't help that her father, who appeared to be a similarly combative character to his daughter, opposed her notion of a career in science. In her second year at Newnham College, Cambridge, he threatened to cut off her fees, urging that she switch to some practical application in support of the war effort. Only when he was dissuaded by her mother and aunt did he relent and allow her to continue her course.

Franklin studied physical chemistry, which involved lectures, extensive reading and laboratory experience in physics, chemistry and the mathematics that applied to these disciplines. One of the mandatory texts she read was Linus Pauling's *The Nature of the Chemical Bond.*

The youthful Rosalind Franklin was disappointed when she ended up with a good second, and not a first, 'bachelor's' degree in 1941. Even then, such was the lingering prejudice against female graduates in science that she was forced to wait in an unseemly uncertainty, one shared with all previous female graduates of Newnham, until her due qualification was formally granted, retro-spectively, in 1947.

Like Francis Crick, Franklin was seconded to National Service during the Second World War, studying the density and porosity of coal for a PhD, in which she helped to classify different types of coal in terms of fuel efficiency. Post-war, she followed this up with a research stint working under the direction of Jacques Mering at the Laboratoire Central des Services Chimique de l'Etat, in Paris. Here, Mering introduced her to the world of X-ray crystallography which he used to study the structure of fibres, such as rayon. 'With his high tartar cheekbones, green eyes and hair combed rakishly over his bald spot', Franklin was surprised to discover that Mering was Jewish, as well as being 'the archetypal seductive Frenchman'. The still youthful, and perhaps naïve, Rosalind Franklin appears to have fallen in love with Mering, who was already married, but whose wife was 'nowhere in evidence'.

Brenda Maddox, one of Franklin's biographers, would draw attention to the fact that Franklin's most imaginative and produc-tive research was conducted when she was teamed up with male scientists of Jewish background. Mering also appeared to be attracted to the trim, slender young woman, with the lustrous dark hair and glowing eyes. They would spend entire days and on into the evenings deep in discussion and argument over likely meanings of X-ray plates and atomic structures.

However, Franklin's infatuation with Mering would be painfully halted when, in January 1951, she took up a post as research associate at King's College London in the Medical Research Council Biophysics Unit, directed by John Turton Randall. Her appointment happened to coincide with a major post-war rebuilding within the department, designed to accommodate new ambitions within the nascent field of biophysics. The precise nature and purpose of her appointment has since become the subject of debate. In part some confusion has arisen because Randall changed the scope of her appointment in between first confirming it and Franklin taking up the post. She had initially agreed to carry out X-ray diffraction studies of proteins, but Randall wrote to her before she took up her appointment, suggesting that she change direction to the study of DNA. According to Maurice Wilkins, this was at his suggestion. Whether at Wilkins' suggestion or Randall's own idea, Franklin agreed. She was offered the assistance of a promising graduate student, Raymond Gosling, to work with. But there was an inherent problem with this new direction.

Wilkins, who was Deputy Director of the MRC Unit based at King's College, was the same scientist who had first lit the fuse of inspiration for Watson in the 1950 Naples lecture. Wilkins had initiated the research into DNA in the department, but happened to be deputising once again for Randall in America at the time of Franklin's appointment. Up to now Gosling had been working with Wilkins on DNA; even after his return from America, Randall failed to inform Wilkins about the terms he now proposed for Franklin's job description. This led to what Franklin's later research colleague, Aaron Klug, would describe as 'an unfortunate ambiguity about the respective positions of Wilkins and Franklin, which later led to dissension between them and about the demarcation of the DNA research at King's'.

This is a short quote from the typed letter from Randall to Franklin, specifying her working conditions:

> . . . *as far as the experimental X-ray effort is concerned there will be at the moment only yourself and Gosling, together with the temporary assistance of a graduate from Syracuse, Mrs. Heller* . . .

While this clearly suggests that Franklin was expected to take on the X-ray diffraction work, the qualification 'at the moment' is too vague to interpret. But there is nothing in this letter to suggest that Franklin should ignore the work performed by Wilkins, or that she should refuse to collaborate with the rest of the department in her approach to the DNA problem.

Wilkins, working with Gosling, had initiated the X-ray diffraction studies on DNA in the department, and in particular obtaining the best resolution diffraction photographs that existed up to this date. They had demonstrated a key property of DNA – that it had a regular, crystal-like molecular structure. In Paris Franklin had learned, and improved upon, X-ray diffraction techniques for dealing with substances of limited order. But even Klug, an ardent supporter of Franklin, admitted that in relation to the work conducted by Franklin in Paris, 'It is important to realise . . . Franklin gained no experience of such formal X-ray crystallography.'

Back in early 1950 Wilkins had complained of poor-quality X-ray apparatus that was not designed for the scrutiny of exquisitely fine fibres. At his suggestion, the department had purchased a new and better-quality X-ray tube to be set up in the basement, but it had lain there for a year or more unused while Wilkins was distracted by the multiple tasks that fell to a busy deputy director of the unit. On her arrival, Franklin, not unnaturally, believed that she was there to take over the DNA work as her personal project. However, the returning Wilkins expected that Franklin had been brought in as his collaborator, to take up the research from where he had already developed it. He would subsequently admit that he was unqualified to take the X-ray diffraction work further and needed exactly such a dedicated and qualified collaborator. 'That's why we hired Rosalind Franklin.'

Unfortunately, Franklin and Wilkins now disagreed as to her role. Even so, rancour was neither necessary nor inevitable between the two scientists, personally or scientifically. These difficulties, provoked by Randall's vagueness, might have been readily overcome with goodwill on both sides, but Franklin, in the opinion of both her biographers, was not inclined to cooperate.

Much has been written about prejudicial attitudes to women in science at this time. In particular an American journalist, and personal friend of Franklin's, Anne Sayre, would write a biography of her in which she suggested that King's College was particularly unfriendly to female scientists, with Franklin struggling to assert her presence in a domain that was almost exclusively male. But when another American journalist, Horace Freeland Judson, looked into this claim, he discovered that of the 31 staff working at King's at this time, eight were female, including some working in a senior position in Franklin's unit. A second biography of Franklin, by Brenda Maddox, confirmed that women were, on the whole, well treated at King's College. Crick made the same point in his biography – and Crick had come to know Franklin well in the years following the DNA discovery. Even in Sayle's more trivial complaint – that the main dining room was exclusively forbidden to women, who were thus precluded from lunchtime conversation – is misleading. There were two dining rooms. One was limited to men, but this, in the main, was used by Anglican trainees. The main dining room, used by the departmental staff, including Randall himself, was open to all.

The frosty relationship between Wilkins and Franklin was not the result of anti-female prejudice – it even seems unlikely to be the result of Randall's peculiar wording in the letter – but it appears to be more directly related to a personality clash between the two scientists. Of the two, only Wilkins ever seems to have made any attempt at compromising. He asked other colleagues what he should do, but Alexander (Alec) Stokes, his closest colleague, was even meeker than he was. In Brenda Maddox's opinion, the two should have got on well; Wilkins was gentle in manner and, despite his lack of self-confidence, was attractive to women. He was mathematically fluent and immersed in the very problems that concerned Franklin. But 'confrontation', in Maddox's words, 'was Franklin's tactic, whenever cornered'. In an earlier confrontation with her professor, R. G. W. Norrish, when working on a postgraduate research project at Cambridge, she would confide, 'When I stood up to him . . . we had a first-class row . . . he has made me despise him so completely I shall be quite impervious to anything he may say in the future. He gave me an immense feeling of superiority in his presence.'

Sayre, who championed her friend, would admit that Franklin's ogrish depiction of her professor was unkind and inaccurate. Professor Norrish was awarded the Nobel Prize in Chemistry in 1967.

Sayre had a correspondence with Norrish in which she described Franklin as 'highly intelligent . . . and eager to make her way in scientific research', but also 'stubborn, difficult to supervise' and, perhaps most tellingly, 'not easy to collaborate with'. In Maddox's opinion, 'If Rosalind had wished, she could have twisted Wilkins around her little finger.' The fact is she had no wish to collaborate with him. This left Wilkins isolated locally so instead he turned to Crick and Watson at Cambridge. It also meant that Franklin was equally isolated. To the commonsensical Crick, this may have been a crucial factor when it came to working out the molecular structure of DNA. 'Our advantage was that we had evolved . . . fruitful methods of collaboration, something that was quite missing in the London group.'

In that same year of Franklin's appointment, just before Wilkins headed for America, he asked his colleague, Alec Stokes – another Cambridge-educated physicist – if he could work out what kind of diffraction pattern a helical molecule of DNA would project onto an X-ray plate. It took Stokes just twenty-four hours to do the mathematics, largely figuring it out while travelling home on the commuter train to Welwyn Garden City. A helical model fitted very closely with the picture Gosling and Wilkins had obtained in their diffraction pictures of DNA. It would appear that if anybody first confirmed that DNA had a helical structure, the credits must surely include Wilkins, Gosling and Stokes – the latter would subsequently lament that, in retrospect, he might have merited 1/5000th of a Nobel Prize.

In November 1951, Wilkins told Watson and Crick that he now had convincing evidence that DNA had a helical structure. Watson had only recently heard Franklin say something similar in a talk about her research during a King's College research meeting. This inspired Watson and Crick to attempt their first tentative three-dimensional model for DNA.

But where to begin?

Taking their cue from Linus Pauling, Watson and Crick decided that they would attempt to construct a three-dimensional physical model of the atoms and molecules that made up DNA with their covalent and hydrogen bond linkages to one another. On the face of it, the structure was made up of a very limited number of different molecules. There were the four nucleotides – guanine, adenine, cytosine and thymine – but they also knew that the structure contained a sugar molecule, deoxyribose, and a phosphate molecule. The phosphate was likely to be playing a structural role, perhaps holding the thread together, much as phosphate is a key structural component of our bony human spine. In the colloquium at King's, attended by Watson, such was his lackadaisical absence of focus that he completely missed the importance of Franklin's statement that the phosphate-sugar 'spines' were on the outside, with the coding nucleotides, the GACT, on the inside. As usual, he had eschewed making notes. All that seemed to intrigue Watson was the fact that the King's people were uninterested in the model-building approach developed, with such aplomb, by Pauling.

In 1952 Franklin appears to have undergone a drastic change of heart in her own thoughts on the structure of DNA. She had in her possession a brilliantly clear X-ray picture of DNA, taken by Gosling, that clearly showed a helical structure to the molecule. She called this her 'wet form', and also her 'B form'. But she had even clearer pictures of a different structure of the same molecule in its 'dry form', or 'A form', that did not appear to suggest a helix. The contrast between the two forms caused Franklin to dither as to whether the DNA molecule was helical. There is a suggestion that she may have asked the opinion of an experienced French colleague, who advised her to place her bets on whichever form gave the clearest pictures. She must have been altogether aware of the advice her ignored colleague, Wilkins, would have given. Unfortunately, she ended up putting the B form into a drawer, meanwhile focusing most of her research over that year into the A form.

Early that same year Watson and Crick made a first attempt at building a triple-stranded helical model of DNA, with a central phosphate-sugar spine. When Wilkins brought Franklin and

Gosling up to Cambridge to view the model, they broke out into laughter. The model was absolute rubbish. It did not fit at all with the X-ray diffraction predictions. Thanks to Watson's lackadaisical focus, and his failure to take notes at Franklin's colloquium, he had made the cardinal error of putting the phosphate-sugar spine at the dead centre of their helix and not on the outside, as Franklin and Gosling had clearly deduced.

Sayre, who rightly defended Franklin from the egregious caricature depicted by Watson's book, loses track of the contribution of Wilkins and Gosling. It is true that Franklin and Gosling had produced some of the clearest pictures yet of the B form of DNA, pictures of such clarity that they did come astonishingly close to the truth of its molecular structure. But then, confused for a year by the two seemingly different patterns of the A and B forms, Franklin veered away from her own earlier conclusions and for a year she took the view that DNA wasn't helical at all. Sayre appears to refute this, but Gosling would subsequently confirm Wilkins' account of how, on Friday 18 July 1952, Franklin goaded Wilkins with an invitation to a wake. The invitation card announced, with regret, the death of the DNA helix (crystalline) following a protracted illness. 'It was hoped that Dr M. H. F. Wilkins would speak in memory of the deceased.' At the time Wilkins assumed it was typical of Gosling's sense of humour. But many years later he would discover that it was Franklin who had written the card, and it confirmed her refutation of any helical structure of DNA in that confused year.

*

In the middle of the same year, 1952, Crick struck up a conversation with a local young Welsh mathematician called John Griffith, whom he met one evening after a talk at the Cavendish by theoretical astronomer Thomas Gold. Gold had captured Crick's imagination with the notion of 'the perfect cosmological principle'. Wondering if there might be some equivalent 'perfect biological principle', Crick pressed Griffith, who was interested in how genes replicated, about the work of an American chemist, Erwin Chargaff, who had discovered that the nucleotides in DNA formed flat linkages with one another. It was curiously reminiscent of Pauling's discovery of how the amino acids that made up the primary chains of proteins, known as 'peptide bonds', also joined up to form flat two-dimensional planes. In Crick's mind it invoked a vague notion that this might be something to do with DNA replicating itself. He asked Griffith if he could work out which of the four nucleotides would pair off with which. Griffith confirmed that the likely pairing was C with G and A with T . . .

But even then the penny did not drop.

Erwin Chargaff was yet another Austrian scientist who fled Europe in the years leading up to World War Two and headed to the US, where he became Professor of Biochemistry at Columbia University. His interest was the study of nucleic acids. We might recall that much of the disbelief around Avery's discovery centred on the fact that geneticists had been misled by the erroneous notion of Levene's 'tetranucleotide hypothesis', which proposed that DNA comprised repeats of the same cluster of four nucleotides. Such a simple formula would be incapable of storing the vast memory demanded for the molecule of heredity, so it was wrongly assumed that DNA could not be the answer to the gene.

Chargaff didn't give a damn what the geneticists thought of Avery: he was deeply impressed by his findings. And if Avery was right, and DNA *was* the molecule of heredity, the DNA sequences of a horse, say, would be different from that of a cat, or a mouse, or a human being. In Chargaff's words, 'There should exist chemically demonstrable differences between [their] deoxyribonucleic acids.' These differences should be demonstrable in the proportions of the four different nucleotides. It might appear that a four-letter code would be limited in delimiting the wide variety of genes that occurred in nature, but if we were to regard the nucleotides as letters in a four-letter alphabet, and the genes as words, the potential for different arrangements of just four 'letters' in words a thousand or more letters long could easily explain the complexity needed for the make-up of genes.

Technology was limited in the late 1940s and early 50s, but Chargaff modified a technique, known as paper chromatography, to read off the different proportions of the four nucleotides in any given sample of DNA.

After four years of laborious experiment, analysing the DNA of yeast, bacteria, oxen, sheep, pigs and humans, Chargaff had his answer: the four nucleotides that made up the word of the gene were not present in the equal proportions one might expect from Levene's hypothesis. For example, human DNA, extracted from a gland in the chest called the thymus, yielded 28 per cent adenine, 19 per cent guanine, 28 per cent thymine and 16 per cent cytosine. The tetranucleotide hypothesis could now be ditched. But Chargaff took it further. He demonstrated that the proportions of the nucleotides varied between species – meanwhile, the proportion of nucleotides was always the same for members of the same species, and indeed was the same from organ to organ and tissue to tissue within the same species. He also noticed something else: by inference, the sum of the molecules of adenine and thymine was very similar to the sum of the molecules of cytosine and guanine.

This was a groundbreaking discovery.

In May 1952, by remarkable happenstance, Chargaff arrived in person at Cambridge University where Kendrew introduced him, over lunch, to Watson and Crick. Chargaff was offended by how little they knew about his work. In his opinion they appeared to know next to nothing about the actual chemistry of nucleotides. In Chargaff's subsequent description to Judson: 'I explained our observations . . . [that] adenine is complementary to thymine, guanine to cytosine.' But as far as Chargaff could see, all that preoccupied Watson and Crick was the race to construct a DNA helix to rival Pauling's alpha helix for proteins. Watson would subsequently recall Chargaff's open scorn for the 'two men who knew so little – and aspired to so much'.

Chargaff was largely right in his assessment of Watson and Crick's ignorance of the biochemistry at the time. Crick knew nothing about Chargaff. No more did he understand that the pairing of the nucleotides involved not the covalent chemical bonding found in stable molecules but the weaker hydrogen bonding. What then was he to make of Chargaff explaining the importance of the one-to-one ratios of cytosine and guanine, and adenine and thymine, in the molecule of DNA?

Crick then had a brainwave: what if this signified a natural chemical attraction between these nucleotides?

Might this not play a vital role when an existing strand of DNA copied itself to a daughter strand? Every C would attract a G, and every A would attract a T in the daughter sequence – and the whole thing would revert to the maternal sequence when the daughter strand replicated in its turn. He took it a stage further. What if DNA comprised two threads, complementing one another in exactly this way? If and when the two threads broke apart and copied themselves, they would create an identical pair of threads, a new identical chain.

It seemed too incredible to be true that the great and profound mystery of heredity might be explained on the basis of these simple chemical couplets, with their specific attractions to one another.

Then Crick and Watson made a mistake, not in scientific terms, but in human terms. They sat back and thought about all that they knew but did nothing about constructing a new model. It almost cost them everything. In December 1952, Peter Pauling, son of Linus, then working as a graduate student at the Cavendish, informed Watson that he had just received a letter from his father to say that he had worked out the structure of DNA. The following month, Peter showed everybody an advance copy of the paper, which was scheduled to be published in February 1953 in the *Proceedings of the National Academy of Sciences*. Watson and Crick would later confess with what sinking hearts they read the paper, which proposed a triple helix with the phosphate-sugar spine at the centre. For a brief interval they were dumbfounded, wondering if their own model, dismissed by Wilkins and Franklin, might have been correct after all. But then they realised that all of the scorn heaped on them by the X-ray crystallographers also applied to Pauling's model. This time it was the great chemist's turn to blunder.

The race was now on again to get it right. Where the Cambridge duo had agreed to stay away from DNA, Watson was now convinced that if they did so, Pauling would beat them all to the prize.

A few days after reading Pauling's paper, Watson took it down to King's College London, where, according to his biography, he talked about it first with Franklin – who, according to Watson, flew into a rage. In Watson's opinion her rage was provoked by his criticism of her rejection of helical structures. But it would also appear that Watson deliberately provoked Franklin into that response. Discovering that . . . 'Rosy was not about to play games with me, I decided to risk a full explosion. Without further hesitation I implied that she was incompetent in interpreting X-ray pictures.'

It was hardly surprising that Franklin flew into a rage.

Much has also been made of the fact that, without consulting

Franklin, Wilkins showed Watson the photographic copy of the particularly clear X-ray picture of the wet form of DNA obtained the previous May – a picture that confirmed without a shadow of a doubt that the molecule of DNA was a helical structure. In fact, Watson, Crick and Wilkins were already long convinced of the helical structure of DNA. Wilkins makes clear in his belated biography, published in 2003, just a year before his death, the X-ray photograph that Watson crowed about was not stolen from Franklin but was passed to him by Gosling, who had taken the photograph in the first place, and who would have assumed, now that Franklin was leaving, that she could have no objection to his doing so. Gosling still needed to complete his PhD thesis without the departing Franklin's supervision, and so had every reason to show his own work to the deputy director of the unit who would now be tutoring him. Gosling himself would confirm that, 'Maurice had a perfect right to that information.' Gosling was clearly fed up with the rancour provoked by Franklin's refusal to collaborate with Wilkins, bemoaning a time when, 'There was so much going on at King's before Rosalind came.'

At the time, Franklin was preparing to leave King's to join the staff at the Biomolecular Research Laboratory at Birkbeck College, London, where she would work under the directorship of J. D. Bernal. To her credit, in her two years at King's Franklin had made a series of original discoveries about DNA. Her research included revealing that DNA existed in two different forms, which she had labelled A and B; that one form could readily turn into the other; and that she had hard proof that the phosphate spine was on the outside. This latter revelation established that it was readily exposed to take up water molecules which wrapped the molecule in a protective sheath within the nuclear environment, keeping it relatively free from the interaction of neighbouring molecules while making stretching of the molecule easier.

Once ensconced at Birkbeck, Franklin appears to have settled into a fruitful and amicable working routine with her boss, Bernal, and graduate student Aaron Klug. Here she ceased to work on DNA fibres and instead focused on the molecular probing of viruses, producing some of her finest work. On her later tragic and untimely death, when she left her worldly possessions to Klug and his family, her scientific obituary was admiringly and respectfully written by Bernal in *The Times* and the scientific journal, *Nature*:

> *Her life is an example of single-minded devotion to scientific research . . . As a scientist Miss Franklin was distinguished by extreme clarity and perfection in everything she undertook. Her photographs are among the most beautiful X-ray photographs of any substance ever taken.*

Neither Franklin nor Wilkins was aware at the time when Watson stormed in with the Pauling paper, that he and Crick were now determined to construct a new model of its three-dimensional structure. Following the triple helix debacle, Bragg had forbidden them doing any more work on DNA. Through all the emotional ballet, recalled so vividly by Watson, we should acknowledge that Watson did keep the King's group informed about his and Crick's thinking, and he had attempted to acquaint Franklin with a relevant publication, coming from a major potential rival. We might also note that up to the final decipherment exercise, Watson, Crick and Wilkins had communicated openly with one another. If Franklin was not privy to these discussions, it was at her choosing. In neither of the biographies of Franklin is there mention of her being inspired by Schrödinger's book or his theory of an aperiodic crystal. She had not chosen DNA as her research theme, it had been suggested to her by Randall – though she evidently saw it as a challenge befitting her growing expertise and fascination with X-ray crystallography.

In their enthusiasm for the model-building approach, Watson and Crick had explained all about it to Wilkins. In passing over the reins of the DNA research to him, they even lent him their jigs for making the necessary parts of the model. But not only had Franklin rebuffed cooperation with Wilkins, the King's group had eschewed the opportunity of taking up the Pauling-inspired modelling approach. And now, at what must have appeared a critical moment in time, what were Crick and Watson to make of the fact that Franklin was leaving King's, abandoning her work on the DNA fibre, and at the same time Wilkins had also stopped working on DNA, waiting, as he confessed, for the dust of Franklin's departure to settle before vaguely starting anew.

Watson had every reason to assume that Pauling, bruised by his own published error of a triple helix structure for DNA, must now be more intensively engaged than ever with the problem – he must surely be formulating a new molecular approach. After the heated encounter at King's, Watson and Wilkins shared a meal and a bottle of Chablis. But their conversation over dinner produced no new inspiration. For Watson the key theoretical difficulty was not whether the DNA molecule was a helix, but whether the helix was a triple or a double chain. Wilkins still favoured three chains over two, but in so far as Watson could tell, Wilkins' reasoning was not foolproof. By the time he had cycled back from the station in Cambridge and climbed over the back gate to make a late return to college, 'I had decided to build two-chain models.' He must have chuckled at a humorous inspiration he would subsequently pass on to Crick the next morning. Francis would just have to agree. 'He knew that important biological objects come in pairs.'

This inspiration – and there appears to be no other word for why Watson decided to focus on a double helix – would prove to be exactly what was required to fit the Chargaff data and Crick's ideas on how DNA might replicate itself. Given the new state of affairs at King's, even Bragg saw the sense of allowing his unruly young scientists to return to the mystery of the gene, most particularly so since it might give his group the advantage of a triumph over his academic rival, Pauling.

The modelling went into overdrive, with Watson putting together scale models of the different chemicals involved in the structure of DNA, the four nucleotides – G, A, C and T – the phosphate molecule and the sugar molecule, deoxyribose. Obstinate in his notion that the spine, which probably involved the phosphate and sugar, had to be internal, Watson attempted to construct a new model with the phosphate-sugar spine still on the inside. But Crick, playing symbiotic devil's advocate, insisted this just did not fit the X-ray data. Indeed, Franklin and Gosling had both insisted that the phosphate spine must be on the outside. Watson now confessed that he had simply refused to take this into account because it made the modelling too easy and introduced an enormous variety of possibilities. But now, persuaded by Crick, he switched to putting the phosphate-sugar spine on the outside – as an exoskeleton, like one sees in the insect world – and then attaching the nucleotides so they projected into the middle of the double helix made by the spiralling phosphate-sugar spines. In spite of Chargaff's work, and in spite of Griffith's advice to Crick, Watson persisted in attempting to attach like with like, for example A to A, G to G. It just didn't work.

In the middle of all this, happenstance again contributed to the story. An American scientist, Jerry Donohue, a former protégé of Pauling's, paid a visit to the Cambridge lab. An expert on hydrogen bonding, Donohue now corrected their models to suit the quantum implications.

Watson and Crick now felt more confident that it had to be a double-stranded helix, with the two strands reading in counter directions – the sense and anti-sense we take for granted today. The two strands had to line up, with the complementary nucleotides linking to one another through hydrogen bonding. Watson sat down at his desk and cut out pieces of stiff cardboard in the shape of the nucleotide molecules, looking at how the actual shapes fitted with one another, hoping to see some pairing possibilities.

Suddenly I became aware that an adenine-thymine pair held together by two hydrogen bonds was identical in shape to a guanine-cytosine pair held together by at least two hydrogen bonds.

Today we recognise that the latter is held together by three hydrogen bonds. If we peer at the basic shapes in the figure below we can see what became obvious to Watson.

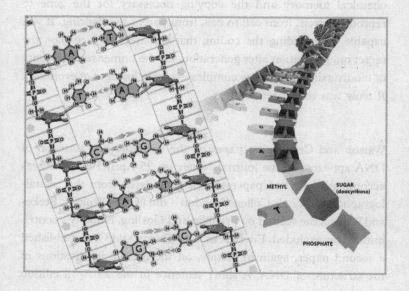

The sceptical Crick, on arrival at the lab to examine Watson's matching shapes of cardboard, almost immediately agreed with Watson's brainwave. The complete model could now be assembled in three-dimensional space – a conflagration of bits of wire, cut to the right lengths to represent covalent and hydrogen bonds, the molecules constructed of the composite atoms, and the whole assemblage suspended by clamps from tall vertical steel rods. The resulting double helix coiled around the central rods, rising in a spectacular conflation of wires and hand-cut, molecular-shaped plates from the lab bench upwards to the ceiling.

Everybody who saw the subsequent model reacted with awe. It was as if in the briefest look at it they saw immediately that it had to be right. It wasn't merely right: it was a spectacularly gorgeous creation – a beauty to behold. All the more so since it was immediately obvious to everybody who saw it that it explained all that was demanded of the mystery of the gene, in terms of chemical memory and the copying necessary for the gene to reproduce itself, from cell to cell, from parent to offspring. It was capable of providing the coding that was necessary to pass the secret on, generation after generation, for the immense complexity of biodiversity, and for the complex evolving lineages of evolution. It truly was the secret of life.

*

Watson and Crick's first paper on the structure and function of DNA appeared in the journal *Nature* on 25 April 1953, and was accompanied by two papers in the same issue from the crystallographers at King's College London – the first by Wilkins, Stokes and Wilson; the second by Franklin and Gosling. Nobody's contribution was excluded. Five weeks later Watson and Crick published a second paper, again in *Nature*, on the genetic implications of the structure of DNA. A short sentence in the 25 April edition would capture the attention of scientists throughout the world: 'It has not escaped our notice that the specific pairing we have postulated immediately suggests a possible copying mechanism for the genetic material.'

Those papers would inevitably change the world of biology, evolutionary biology and medicine. Indeed, the ramifications are still echoing through our world, and penetrating much wider and deeper into society than Watson and Crick could have possibly imagined.

It is extraordinary to realise that, less than two years after the start of their ad hoc partnership, Watson and Crick had correctly figured out the three-dimensional chemical structure of DNA. Crick was 37 and still had not completed his PhD, and Watson was still a postgraduate student, aged just 25. At a superficial glance, there would appear to be no logical reason why these two seeming misfits should be the discoverers of the three-dimensional chemical structure of DNA. They had done little or no lab work in the prelude to the discovery. They were lowly in their positions within the lab – Crick a research assistant and Watson a graduate student. They were impecunious, living in impoverished surroundings, yet uncaring about all that. They had only belatedly realised the relevance of discoveries made by other scientific contributors. Their official duties had nothing to do with DNA. Crick was still trying to complete his PhD thesis on the X-ray diffraction of polypeptides and proteins while Watson was supposed to be helping Kendrew to crystallise the molecule of myoglobin. The head of their department, Sir Lawrence Bragg, was, through most of their efforts, opposed to their work on DNA. In the manner in which science normally works, the pair should never have made their discovery. There were colleagues, like Willy Seeds who insulted Watson at the foot of the Swiss glacier, who thought that Watson in particular didn't deserve the acclaim. But the fact is, they both did.

The detractors are missing the point: what Watson and Crick achieved was an act of sublime creativity, like the plays of Shakespeare, Da Vinci's Mona Lisa, or Beethoven's Ninth Symphony. Admittedly this was not artistic creativity. Rather, like Newton's discovery of gravity, Darwin's discovery of natural selection and Einstein's Theory of Relativity, this was an act of scientific creativity that opened a new window of understanding onto our understanding of Life itself – and, at the most profound of levels, what it is to be human.

In 1962, Crick, Watson and Wilkins shared the Nobel Prize in Physiology or Medicine for the discovery of the structure of DNA. The only one of the three to mention Rosalind Franklin was Wilkins, who also acknowledged Alexander Stokes' 1/5000th contribution. Tragically, Franklin had died of ovarian cancer some four years earlier, at a time when her work on viruses was becoming globally recognised as among the finest achievements in X-ray crystallography. Some, including Sayle, have queried if Franklin might have taken Wilkins' place on the rostrum had she lived. It's a moot question, but I personally think it unlikely. Wilkins initiated the DNA research at King's, inspired, like Watson and Crick, by Schrödinger's book. His X-ray diffraction picture – actually taken by Gosling – inspired Watson's arrival into the Cambridge lab. His cooperation with Watson and Crick was so close and formative in the discovery that Watson wanted to include his name in the famous first paper. It was Wilkins' modesty and integrity that caused him to refuse the honour. This is why I doubt that Franklin would have replaced Wilkins on the rostrum in 1962. But I do believe that there might have been a second, rather more likely, opportunity for recognition of the contribution of Rosalind Franklin to X-ray crystallography, one that is suggested in the great admiration felt for her work by such an eminent figure as Bernal.

When she moved to Birkbeck College Franklin found a happy

working relationship with a Lithuanian Jewish chemist and biophysicist, Aaron Klug, who, following graduation in South Africa, had
arrived in the UK on a research fellowship to complete his doctorate
in X-ray crystallography at Trinity College, Cambridge, in 1953.
This was, of course, the year of publication of the DNA discovery.
At Birkbeck Franklin took Klug under her wing, forming a close
working relationship and friendship that would continue for the
rest of her life. We know that after Franklin's untimely death Klug
took her techniques further to be rewarded with the Nobel Prize
in Chemistry in 1982. The official declaration of his meriting the
Prize was: 'for his development of crystallographic electron micros-
copy and his structural elucidation of biologically important acid-
protein complexes'. How likely is it that, had Rosalind Franklin
lived, she would have shared the podium with Aaron Klug for
their cooperative effort?

*

Some nine years earlier, on 12 August 1953, five months after
Crick and Watson had first modelled the double helix, Francis
Crick wrote a letter to Erwin Schrödinger in which he thanked
him for the inspiration of his book. In the letter he described how,
in the structure of DNA, they had indeed discovered the 'aperiodic
crystal' that he had predicted would be the molecular code for
life.

six

The Sister Molecule

I have the feeling that if your structure is true, and if its suggestions concerning the nature of replication have any validity at all, then all hell will break loose, and theoretical biology will enter into a most tumultuous phase.

MAX DELBRÜCK, WRITING TO WATSON

Judson, widely recognised as the historian of the DNA story, described the elucidation of its structure as 'a siege, a conquest'. Given the discovery of the three-dimensional structure of DNA, with its four-letter code for the storing of heredity, one might have anticipated enlightenment, but instead the prevailing atmosphere was one of confusion. Watson and Crick's discovery had provoked a storm of new questions. To begin with, was DNA the answer to the coding of heredity to all of life? At least this question was already answered; Avery had set the ball rolling by discovering it in bacteria. The phage school worked on it in viruses. Chargaff had confirmed the same in a range of different life forms. DNA was universal. The next major question was this: how did its incredibly simple four-letter code – G, A, C and T – translate into the complexity of the estimated 80–100,000 proteins that were essential for the structure and functioning of our human bodies, and the bodies of every other living creature?

Crick would later recall that they already had an outline of the answer to the protein enigma. Since the spine of the helix was made up of sugar and phosphate repeats, the only chemicals capable of coding for heredity, and translating to proteins, were the four nucleotides – otherwise known as bases, or base sequences – GACT. Some advances had already been made into this mystery. Thanks to the pioneering evolutionary biologist Thomas Hunt Morgan, working in his fruit fly laboratory at Columbia University, we knew that the genome was divided into chromosomes. Thanks to Morgan, Muller and others, we knew that the chromosomes were themselves parcelled into discrete sections, called genes. A further step, that a gene coded for a specific protein, was first postulated by a British doctor, Archibald E. Garrod, as early as 1908, when he figured out that the inherited illness known as alkaptonuria was probably the result of a defective enzyme. An enzyme is a protein that speeds up the rate of a chemical reaction in living systems. But Garrod couldn't take it that vital step further and prove that the defective enzyme was the expression of a defective gene. The essential link between genes and proteins was confirmed by two Americans – a geneticist, George W. Beadle, and a biochemist, Edward L. Tatum – who were working on the heredity of eye colour in fruit flies. By 1941, shifting their focus to a fungus that infected mouldy bread, they had showed that a single gene coded for a specific enzyme involved in the mould's living chemistry. It was this discovery that resulted in the maxim: 'one gene one protein'. But how did the four-letter DNA code of the gene translate to the 20-letter amino acid code of the protein?

For Francis Crick this was the very enigma that had inspired his own 'mad pursuit' after reading Schrödinger's book. Following the discovery of the double helix, Watson would soon be forced to return to the United States, having run out of funding. But Crick would continue to investigate the extrapolation to proteins.

Since the DNA was contained in the nucleus of the cell and protein manufacture always took place in the region outside the nucleus, known as the cytoplasm, the code of the gene had to be copied in some way that allowed it to be sent out of the nucleus and into the cytoplasm. This made Crick consider a sister molecule of DNA known as ribonucleic acid, or RNA.

There are obvious similarities between the two molecules. Both are nucleic acids, made up of varying sequences of four nucleotides. Where DNA is made up of guanine, adenine, cytosine and thymine, or GACT, RNA is made up of guanine, adenine, cytosine and uracil – GACU. RNA also differs from DNA in the fact that it is not a double-stranded helix but, at least in most of its roles, is single-stranded. It also differs from DNA in its sugar, which is ribose where DNA's sugar is deoxyribose. At the time of Watson and Crick's discovery of the 3-D structure of DNA, molecular biologists and geneticists were also becoming increasingly interested in this sister molecule, RNA. Immediately prior to Watson and Crick's monumental discovery, many scientists were beginning to think that RNA must be playing an important role in the way living cells worked.

At the same time there was something elusive about RNA. Where the amount of DNA in different cells of the body, say a brain cell or a liver cell, was always the same, the amount of RNA seemed to vary. To add to the confusion, DNA was only found in the nucleus, meanwhile RNA was found in both the nucleus and the non-nuclear territory, known as the cytoplasm – the part of the cell where most of the active biological chemistry takes place. To confuse things even more, the amount of RNA in a given cell also varied, depending on how active the cell was. A growing cell, or a cell that was producing lots of new protein, had more RNA than a cell that was mature and chemically quiet. Liver cells, for example, which were thought of as the factory for making proteins, were packed with RNA. Moreover, RNA was also found in the same parts of the cytoplasm – the small round bodies known as ribosomes – where protein was manufactured.

It was now becoming increasingly likely that if DNA was the stored code for heredity, and somehow was translated into the amino acid sequences that made up proteins, RNA had something to do with the actual manufacture of those translated proteins. It was easy to see how a stretch of DNA could make an RNA copy – all it would take is for the thymine, or T, of DNA to be replaced by uracil, or U, during the copying. As early as 1947, two Strasbourg-based scientists, André Boivin and Roger Vendrely, had already proposed that the GACT-based sequences of a DNA gene would be copied in this way to the GACU-based RNA messenger, which would ferry the coding out into the cytoplasm where the corresponding proteins would be manufactured at the ribosomes. All that was left was to figure out how the four letters of GACU could translate to the 20-letter amino acid code of the proteins.

Crick was intrigued by a letter from a Russian theoretical physicist, George Gamow, which arrived out of the blue in the summer of 1953, soon after Crick and Watson had published their first iconoclastic paper on the structure of DNA. Gamow, who was part of the group who had come up with the 'Big Bang' theory for the origin of the Universe, had been intrigued by the double helix. In his letter he proposed a way in which the four-letter code of DNA might translate to the amino acid primary sequences of proteins: triplets of the four nucleotides – G, A, C and T – might code for single amino acids. But to Crick this didn't add up. Sixty-four different triplet assortments would result from the random mixing derived from four coding letters, whereas there were only 20 amino acids found in natural proteins. Gamow had thought this through, proposing an ingenious overlap of the triplets, so that what coded for one amino acid might in part code for another. Crick didn't buy it, but he and Watson took Gamow's letter with them when they retired, as usual, to the Eagle for lunch. If nothing else, Gamow's intervention provoked the DNA pioneers into a renewed debate on how the DNA to protein mystery might be cracked.

There would be little further opportunity for them to swap ideas that year once Watson left Cambridge for America. In fact, there would be little progress in the mystery for a number of years.

In the summer of 1954, Crick and Watson teamed up again for three weeks at Woods Hole, Massachusetts. Gamow and his wife were there. Most afternoons Crick and Watson would join the Gamows down by the water's edge, watching the Russian physicist do card tricks and chatting about the same mystery. In the interim, since writing the letter, Gamow had collected together the names of a number of people interested in solving the problem. Somehow or other, and it was likely that both Watson and Delbrück were at the heart of the joke, a 'whisky, twisty RNA party' was called for, with invitations presumably addressed to parties interested in the enigma. This became the inspiration for what came to be called the RNA Tie Club, limited to 20 members, to parallel the number of amino acids. In addition to Crick, Watson and Gamow, the club members included Martynas Ycas, Alex Rich and an Oxford-educated South African, Sydney Brenner. In the spring of 1953, Brenner had been one of a party of young scientists who had driven over from Oxford to Cambridge to see Watson and Crick's 3-D model of DNA. At the time, Brenner was conducting a bacteriophage project for his PhD in molecular biology. In a garden stroll with Watson, Brenner had learnt about the Hershey–Chase experiment. These days he was working as a post-doc at the Medical Research Council's Molecular Biology Laboratory in Cambridge, but he had maintained that early interest in DNA and genetics. Each member of the club received a tie, made to Gamow's design by a haberdasher in Los Angeles. The ties in turn were pinned by individually designed short forms of each specific amino acid – Crick's pin read 'tyr' for tyrosine. It was all a fantasy; the club never met, but, like the phage group, it acted as a rallying focus for the group, who would circulate any papers or news of common interest to members. In the words of British journalist and author, Matt Ridley, who wrote a biography of Francis Crick, the English scientist was 'the dominant theoretical thinker . . . the conductor of the scientific orchestra'.

Brenner showed, mathematically, that the overlapping triplet idea was a non-starter. Crick and Leslie Orgel were joined by Crick's friend and collaborator, the young Welsh mathematician, John Griffith, who tried his hand at ruling out specific triplets of the four letters that simply would not work. For example, they ruled out AAA because it would cause confusion if positioned next to another identical letter, A. By excluding triplets that would cause confusion, they calculated that it only left 20 'sense' permutations. This was duly published as a paper in the *Proceedings of the National Academy of Sciences*, in 1957. Unfortunately, it was utterly wrong.

Nevertheless, by now some useful ideas were beginning to emerge.

A gene, with its long thread-like molecule, made up of a specific sequence of G, A, C and T, often a thousand or more letters long, coded for a specific protein whose primary structure was again a long thread, made up of the 20 amino acids in a very specific sequence. They even knew by now that sickle-cell disease, in which there was an abnormal oxygen-carrying haemoglobin in the red cells, was caused by a mutation in the gene that coded for the beta globin. The mutation in the gene had translated into a faulty coding for the haemoglobin protein. Crick now focused on ideas that were coalescing from several different quarters, which boiled down to the fact that there were likely to be two quite different forms of RNA involved in the translation from nuclear-based DNA genes to the coding for protein assembly in the ribosomes. One form – now called messenger RNA, or mRNA – copied the code from the entire gene in the chromosomes within the nucleus and carried the code out of the nucleus to the ribosomes. Interestingly, messenger RNA was found by a group of researchers at Harvard, working in Watson's new laboratory there. Meanwhile, a second form of RNA, called transport RNA, or tRNA, picked up single amino acids and, guided by the coding carried on the messenger RNA, added the right amino acids, one by one, to the assembling protein chain. In this way the nucleic acid coding of the gene translated to, and was ferried into, the ribosomes to construct the corresponding protein chain.

The coding triplets were eventually discovered through trial and error by scientists including Marshall Nirenberg, Gobind Khorana and Severo Ochoa. Today, we know that triplets of DNA, now called 'codons', code for specific amino acids, but a single amino acid can have more than one corresponding codon. For example, the amino acid leucine has six different triplet codons: CTT, CTC, CTA, CTG, TTA and TTG; phenylalanine has two, TTT and TTC; while methionine has just one, ATG. Moreover, there are specific triple permutations – TAA, TAG and TGA – that code not for amino acids at all but for the genetic equivalent of a full stop. These halt the production of a protein at the full stop, and so are known as 'stop codons'.

This was another major step in understanding, but, once again, it begged new questions. The factory-like mechanisms of protein production had to be controlled. How did a cell decide which protein to make? How did it decide when to make the specific protein in the life of the cell? How did it turn protein production on and off?

*

We might recall the group who had earlier contributed to the Watson–Crick discovery (a world-based cooperative of scientists working with the viruses that infect bacteria), the phage group. A trio of Paris-based scientists, André Michel Lwoff, Jacques Monod and François Jacob, were conducting research on phages and their host bacteria at the Pasteur Institute. They focused on the bacterium that was the host for all of the phage experiments, a bug called *Eschurichia coli* – *E. coli* for short – which is the bacterium most commonly found in the human intestine. What interested them, to begin with, was a discovery made by their American colleagues Joshua Lederberg and Edward Tatum, which suggested that, contrary to the prevailing ideas, bacteria had a kind of sex life. Normally bacteria reproduce asexually through a daughter bacterium simply budding off the maternal strain – rather like one sausage being squeezed off in the middle to form two – but now and then a bacterium would fashion a penis-like extrusion through which it would inject its genetic material into another bacterium. The scientists jokily referred to it as 'coitus'.

In 1955, Jacob, working with a colleague called Élie Wollman, explored the way in which genetic information was passed on from one bacterium to another. Realising that bacteria had genes made up of DNA, just like all other life forms, they also knew that the bacterial genes were threaded along a single lengthy chromosome that took the form of a ring, which had a point of attachment to the inside of the bacterial wall. Jacob and Wollman now discovered that during coitus the chromosome was very slowly extruded from the 'male' and through the cell wall into the body of the 'female' bacterium. While bacterial reproduction by budding took only twenty minutes, bacterial sex lasted for roughly two hours. This allowed Jacob and Wollman to conduct a series of 'coitus interruptus' experiments in which they halted the process at timed intervals along the two-hour process. Since the bacterial chromosome always came through in the same order of genes, they could, through looking for the effects of specific mutated genes, plot where along the course of the bacterial chromosome the genes for various different properties were located.

But now the French scientists took the experiment a step further; they set out to discover how those genes were controlled within the bacterium.

They focused on three genes that allowed the bacterium to transport the sugar, lactose, into the body of the bacterium and there digest it into its two smaller component sugars, glucose and galactose. It would be wasteful for the bacterium to activate these genes all the time, even when there was no lactose in its environment. What they discovered was that the genetic chemistry operated a system of control. When there is no lactose around, this triggered the activation of a 'repressor', which halted the production of the three relevant genes. When lactose was present, the repressor was removed and a genetic area alongside the genes, known as the 'promoter', activated the expression of the genes.

We don't need to worry about the precise genetic details. All we need to grasp is that there are regulatory systems that switch genes on and off in every life form. Moreover, these systems have ways of detecting key signals coming from outside the genome – in the above case they are capable of detecting the presence of the sugar, lactose, in the bacterial environment. This was the first scientific demonstration of what we now call genetic 'regulatory' control. It would result in Lwoff, Monod and Jacob sharing the Nobel Prize in Physiology or Medicine in 1965.

*

The time has come to introduce a little magic. What I have in mind is a maiden voyage in a mystery train. Imagine that we have shrunk to ultramicroscopic size – a thousand times smaller than a retrovirus, so that a human cell would appear the size of a major city and where the individual nucleotides that make up DNA are easily discernible. We can, in the blink of an eye, climb aboard the most exciting part of it, the chugging engine.

With a toot on the whistle, we are off. Up ahead we see a glowing spiralling shape, a spectacularly beautiful double helix, spinning away through the ether from left to right. As we approach the double helix flattens out, still glowing, still running across the dream-like landscape in horizontal fashion, from left to right. We now see that it takes the form of a railway line, with twin rails spaced by closely set horizontally placed sleepers. For a dizzy moment or two, we gaze on the extraordinary structure of DNA from this close.. Then I slow the engine down to a halt. We are now hovering in a steam-filled stillness immediately above the railway line. We hop out so we can take a good look at where we are.

We take a short stroll along the glowing DNA molecule, in the direction that the now-stationary train is pointing.

What we took to be rails are actually banded structures, made up of alternating four-pointed stars and pentagons at right angles to the sleepers. The sheer gorgeousness of it is overwhelming. The stars and pentagons are made up of glowing spheres connected by lines of force.

'So,' you gaze a little closer, in what I imagine to be the same wonder that I am feeling, 'the spheres are the atoms that make up the component molecules?'

'Yes.'

'The crosses and pentagons . . . ?'

'The pentagons are the deoxyribose sugars. The stars are the supporting phosphate molecules.'

'Between them they make the rails?'

'The phosphate stars make up the external spine that Watson and Crick argued about. Each sugar connects the phosphate spine to a sleeper.'

'The lines of force between the atoms are the stable covalent bonds?'

'Yes. The phosphates hold the whole thing together. The sugars are the connection between the spine and the sleepers. Time, perhaps, to take a closer look at the sleepers.'

I allow you the leisure of a stroll along the track, examining sleeper after sleeper.

'The sleepers are attached to the inner angles of every sugar pentagon?'

'Take a closer look at them . . .'

'There are two shapes, joined together in the middle.'

'Two complementary nucleotides, yes – but the join is not exactly in the middle.'

'It has to be a trifle eccentric since the complementary nucleotides are unequal. This join here is closest to the upper rail. In the following sleeper it is closer to the lower rail.'

'The purines, guanine and adenine – G and A – are wider because they contain two contiguous atomic rings. The pyrimidines, thymine and cytosine are shorter because they only contain a single ring.'

'So, one way or another, the sleeper is always made up of a purine and a pyrimidine?'

'Yes. It has everything to do with shapes. Take a good look at the junction in the middle of the sleepers. Look at how the shapes of the nucleotides meet. Does it remind you of anything?'

'It's like the meeting of two pieces of a jigsaw puzzle.'

'Exactly.'

'So that's why they are complementary?'

'Absolutely. And now you know why the molecule has to be constructed exactly as it is.'

'So the real DNA – the nucleotides – is like beads on the string of phosphates and sugars?'

'No. Another scientist, I think a mathematician, said exactly that to Crick. But he was wrong. Crick told him that DNA was itself the string.'

'The DNA has to include the phosphates and sugars, as well as the nucleotides?'

'The construction has to be the whole thing, exactly as it is. Can you see why?'

You take another short stroll, getting the hang of this idea. 'So, the nucleotides, the bases, don't make contact along the thread?'

'Their only meeting point is one-to-one within the sleepers. And always with their complementary partner, A to T, and G to C, or vice versa.'

You gaze down at this wonder, blinking for a moment or two. 'So the code lies in the sleepers?'

'That's right. And the sleepers also explain how the code replicates to form a new daughter strand of DNA. They also explain how code of protein-coding genes translates to proteins. What you need to grasp is the code is contained in just a single rail. In this case if we take the uppermost of the two rails, the code is in the sequence of the uppermost portions of the sleepers. You can read it off if you stroll along the rail and name each nucleotide as you come to them, like a series of letters.'

'I'm reading them: A, A, C, T, G, C . . . I think I'm getting the picture. But why then is there a second rail?'

'The code has already copied itself to a daughter thread. What you see on the opposite rail is this copy.'

'Ah! So – the double helix is actually two copies of the coding DNA?'

'Yes, two complementary sequences. Would you like to see it copy itself?'

'I'd love to see that.'

We stand back and our engine evaporates. The line begins to vibrate.

'What's happening?'

'To copy itself the double helix must part into its two component halves. This normally happens through the action of an enzyme, but it can be done just by heating the system up. Heat adds enough random energy to break open the bonds within the sleepers.'

'So those bonds holding the sleepers together are not stable?'

'No. These are the relatively weak hydrogen bonds we came across when talking about Linus Pauling and his study of chemical bonds.'

As we watch, the sleepers come apart, like pieces of a jigsaw separating. A cloud-like mass appears out of the distance and it begins to move over the now-separated upper rail, with its exposed half sleepers.

'What's that?'

'The cloud is an enzyme – a protein called a synthetase that helps DNA to replicate.'

We watch as the cloud moves along the detached rail, from east to west. It appears to discover the nucleotides it needs from the teeming background, and as it passes along it attaches the complementary nucleotides, A to T, C to G, T to A, G to C, then some other element in the cloud, perhaps another enzyme, or enzymes, grabs the necessary phosphates and sugars to make up the spine.

You're too dazzled by the speed at which the cloud is shuttling along to say a word. In what seems no more than a few moments, the hive of activity has long passed us by and the new twin track is there before us and gleaming into the distance.

'That's it?'

'Almost. I have one more point to make before we' head for home. First we need to take a journey along this new stretch of track.'

In the twinkling of an eye our puffing engine has reappeared before us, ready to roll, and we hop on board, tootle the whistle and head eastwards at a rattling rate.

'Keep your eyes peeled for a red light up ahead.'

After what could have been quite a few miles, you spot it. A pinpoint, red glow in the distance. 'It's in the track, to our right.'

'Yes. It would have to be in the new daughter copy.' I explain that the track closest to us is called the 'sense' track, because it is the original. The daughter copy on the other rail is the 'anti-sense' track. The genetic machinery reads this while travelling in the opposite direction. I slow the engine to a halt so we can see what the red light indicates. 'Look at the sleepers.'

You get down onto your haunches to have a closer look. At first you can't make out anything wrong. The split in the sleeper appears to be as before, with the short section to the left, then the short section to the right. Then you gasp. 'The one on the left is a C, so the other half of the sleeper should have been a G. But it isn't. It's an A.'

'So?'

'The copying mechanism has made a mistake.'

'Yes.'

'So this is . . . a mutation?'

'Yes, it is. To be precise, it's what is called a point mutation, which means a single nucleotide has been mis-copied. But if and when the anti-sense strand copies itself, the mutated nucleotide will attract a thymine to attach to it – in other words the mutation will now be fixed into the double helix. The mutation will, in this way, perpetuate itself. If this were to happen during the formation of the germ cells, the sperm or the ovum, that germ cell would carry the mutation into the genome of the new generation.'

'Are these mutations common?'

'Much more common than one might think. But thankfully there are compensating mechanisms in that moving cloud that will usually recognise and correct them. But mutations do get through from time to time.'

'And this will cause disease?'

'Most mutations don't cause disease. They only do so if they affect a part of the DNA that fulfils an important role in the offspring's internal genetics, or they seriously affect the coding of a protein-coding gene.'

*

In the early years of the twentieth century a Dutch botanist, Hugo de Vries, made the conceptual breakthrough that Mendel's discrete packages of hereditary information could be changed by mutations. Amazingly he did so before we knew anything about the actual structure of DNA or what constituted a gene. As we have just witnessed, a mutation is an error in the nucleotide sequence made during the copying of DNA. Mutations can arise, albeit rarely, during the normal process of DNA copying. They can be induced at a much higher rate if the DNA replication is damaged by external influences, such as exposure to toxic chemicals or excessive doses of radiation.

There are many different types of mutation. What we have witnessed is one of the simplest, in which a single nucleotide has been substituted – a point mutation. A so-called 'frame-shift mutation' would result from the simple deletion of a nucleotide. If we imagine what this would do to the succeeding triple codons we realise that it would interrupt all the triplet sequences following it, and thus would play havoc with the translated protein. Even a point mutation in a protein-coding gene might result in a changed amino acid in the coded-for protein. This is what causes sickle cell anaemia. In this case the mutation replaces what should be an adenine in the beta-globin gene for a thymine. When this translates to the beta-globin protein, the amino acid glutamic acid is replaced by valine. This makes the abnormal haemoglobin in the red cells that causes the disease. If the offspring gets just one copy of the mutated gene, they suffer a milder form of the disease that incidentally protects them from malaria. If they get a double dose of the mutated gene, in other words if they get a mutated gene from both their parents, they get a severe form of the disease that can be fatal early in life. Mutations affecting the cells of tissues and organs in the body, as opposed to the germ cell, are an integral part of the underlying causes of many different forms of cancer.

There are a few additional terms I need to explain to present an outline of basic genetics. Other than the sex chromosomes, X and Y, we inherit 22 non-sex-connected chromosomes from each of our parents. These are called 'autosomes'. This means that we all, both males and females, inherit two copies of every gene that is found on the autosomal chromosomes, which amounts to the bulk of our genes. When a mutation affects an autosomal gene during the formation of the ovum or sperm it will only affect one of the two copies in the offspring. If the remaining normal copy of the gene is enough to supply the body's biochemical needs, there will be no upset in the internal chemistry – no clinical disease. This type is called a 'recessive' gene mutation. But sometimes just one bad gene is enough to give rise to serious disturbance in the internal chemistry, despite the fact the other gene is normal. This is called a 'dominant' gene mutation. When a mutation, whether dominant or recessive, gives rise to a disease this is referred to by doctors as 'an inherited disorder of metabolism' or 'an inborn error of metabolism'.

Many medical conditions arise from dominant genes, for example, Huntington's disease, a condition in which the affected person may develop a progressive cerebral deterioration later in life. The inheritance of one recessive gene isn't enough to cause an inherited disease of metabolism, but if both parents are carrying one copy of the same recessive mutated gene, then there is a one in four chance of the offspring being unlucky enough to inherit mutated versions of the gene from both parents. Since there is no normal copy, this will then give rise to disease.

One in every 2,500 babies born to Caucasian parents suffers from cystic fibrosis, making it one of the commonest of hereditary diseases. It is caused by a variety of mutations affecting a regulator gene, which is known as the cystic fibrosis transmembrane regulator gene, or CFTR, located in the region q31–32 of human chromosome 7, and which codes for an ion channel involved in transport across membranes. Cystic fibrosis is perhaps the most familiar example of an autosomal recessive condition. There are many other recessive genetic disorders that might potentially be cured by the addition of a single 'normal' gene, and these conditions, including cystic fibrosis, are the subject of intensive current investigation aimed at 'gene therapy'.

Another pattern of mutation gives rise to a sex-associated recessive condition. Females have two of the sex-associated chromosomes called 'X' chromosomes, while males only have one X, always inherited from the mother. This means that a recessive gene that happens to be carried on the X chromosome will usually have no serious effects in females but it will behave like a dominant gene if inherited by a male. A sex-linked recessive mutant gene is the cause of haemophilia, a condition that ravaged some of the royal houses of Europe. It is also the cause of the red-green colour blindness that affects between 7 per cent and 10 per cent of men, as well as several types of muscular dystrophy.

Such single-gene mutations will usually be inherited along Mendelian lines, such as the dominantly inherited achondroplasia and Huntington's disease, the recessively inherited cystic fibrosis, and the sex-chromosome-linked disorders. To date, geneticists have identified more than 5,000 single-gene disorders in humans caused by mutations. Some mutations can change the number of chromosomes, as in Down's syndrome, or delete, duplicate, fragment, or otherwise damage the structure of chromosomes, giving rise to a variety of syndromes. As mentioned above, mutation is also a common feature of cancers, which usually arise in fully developed tissues long after embryogenesis. Other chromosomal abnormalities affect the germ cells, where they give rise to a wide range of disorders including aberrant embryological development, with resultant congenital abnormality, as well as a great many inborn errors of metabolism. In all such cases, a clear understanding of the genetic cause, or causes, is the basis for medical prevention and therapy.

The medical approach to mutation includes genetic counselling, for example enabling couples at risk of particular disorders to have essential information so they can make their own decisions on matters of reproduction, and public education about the risks of increasing maternal age, avoidance of risk factors such as irradiation of the germ cells and foetus, caution with respect to drug and chemical exposure, such as thalidomide, and vaccination against rubella. Newer measures, such as preimplantation genetic diagnosis, involve the genetic screening of the foetus at the 16- or 32-cell stages, followed by the selection and implantation of healthy embryos. This is only suitable for a genetic abnormality that is predictable, and the availability of a suitable screening test in isolated embryological cells. It not only reduces the risk of severe abnormality in children in very high-risk circumstances but also removes the mutation, and thus the risk pedigree, in future generations. There are, of course, important ethical and moral principles involved in such therapy for both doctor and patients in what essentially amounts to a positive form of eugenics.

Cancer is another arena in which intensive study of the mutated genes offers the hope of developing more efficient therapies. Here the genetic abnormalities are more complex than in the inherited diseases and very often involve multiple mutations as well as important links to environmental factors. At genetic level, cancer involves a series of steps that involve multiple mutations that deregulate regulatory pathways. New lines of research suggest that these mutations must cooperate with each other for the cancer to develop, so that research aimed at determining the precise nature of the cooperating mutations and the regulatory pathways they affect is a major challenge. The decoding of the human genome has highlighted the genetic alterations that underlie cancers in such unprecedented detail that it has led two American oncologists, Vogelstein and Kinzler, to declare that 'cancer is, in essence, a genetic disease'.

Some 15 to 20 per cent of women with breast cancer have a family history of the condition and 5 per cent of all breast cancers have been linked to mutations in the genes BRCA1 and BRCA2. Geneticists can further predict that women who carry these mutations have an 80 per cent risk of developing breast cancer during their lifetime, so that there are various options that help to reduce the risk, including prophylactic ovariectomy, regular breast screening and the potential of pre-emptive surgery.

In 2006, a systematic multi-centre American study pioneered the screening of more than 13,000 genes taken from human breast and colon cancer cells. Given the 'normal' human genome, they were in a position to compare the genes they found in the two cancers with the normal, revealing that individual tumours accumulate an average of 90 mutant genes. It seems that a much smaller number of these actually play a part in the cancer process, in their estimation perhaps 11 mutations for each of breast and colon cancer. Encouraged by these findings, the US National Institutes of Health is drawing up an atlas of cancer genomes – The Cancer Genome Atlas Project, or TCGA. The aim is to decode the genomes of every human cancer and, by comparing these to the normal, extrapolate the genetic abnormalities that underlie all cancers. A pilot study has begun with cancers of the lung, brain and ovaries. This is far from pie in the sky research; already cancer is being forced back on many different fronts and today some forms of cancers are eminently treatable by surgery, focused radiotherapy and chemotherapy or immunotherapy, so that what might formerly have been a death sentence has become more a chronic but controllable ailment.

seven

The Logical Next Step

Of the three main activities involved in scientific research – thinking, talking, and doing – I much prefer the last and am probably best at it. I am all right at the thinking, but not much good at the talking.

FREDERICK SANGER

In the late 1960s I was privileged to be a medical student at the University of Sheffield. Watson and Crick were still relatively young men, their discovery having been made just fifteen or sixteen years earlier. I can remember my own sense of wonder as our teachers explained the structure of DNA, and the elegance with which its four-letter code translated to proteins. We had lectures on genetics in which we learnt how mutation was a major step in our under-standing of many different hereditary diseases, including the so-called 'inherited errors of metabolism'. We also had lectures on the importance of the same discoveries to the sister discipline of evolutionary biology. I can recall the prevailing sense of excitement that came with the feeling that the biological and medical sciences were entering a new paradigm, based on the growing understanding of DNA and its molecular extrapolations, an understanding that

112

clearly had implications not merely for biological scientists and doctors, but for all of humanity. But at that stage many important questions remained to be answered.

One very obvious question was how did the fertilised egg, or 'zygote', develop into the complex wonder of a human baby? How could this extraordinary chemical, DNA, store not only the heredity of the individual but also the instructional blueprint that was necessary for the single cell of the zygote to give rise to the developing embryo, with its wide range of different cells and tissues and organs that went into making the future baby?

While much was known about the tissue changes within the embryo, little was actually known of the relevant genetics at this time. The work of the scientists at the Pasteur Institute in France offered us the first glimpse into this quandary: they had pioneered our understanding of how a gene is activated by switching on its 'promoter' sequence, and how it is inactivated by switching off the promoter. This was the first step towards what we now call genetic 'regulation'.

Back then we also knew that the cells that make up the different tissues and organs in the human body, such as the brain cells, or the white cells that fight off infection in the circulating blood, or the cells that make up the kidney, or liver, heart or lung, all contained exactly the same DNA in their nuclei. The differences in structure and function between these cells, and thus the make-up of the various tissues and organs, must somehow involve differences in the expression of genes. This provoked two new questions: was the difference brought about by specific genes that were only switched on in specific organs, or was it brought about by different profiles and timing of the expression of the same genes?

The questions did not stop there.

Whatever the explanation, whether special genes for particular cells, or different profiles of expression of the same genes, there had to be a system that decided what gene, or what profile, would be expressed in the different cells, tissues and organs. This must be a key element in the planning and regulation of the developing human embryo – and very likely there would be very similar patterns of regulation of embryogenesis in all animals – and maybe plants as well.

We might recall here Sydney Brenner, who came to work with Crick at the Cavendish Laboratory on the translation of genes to proteins. In 1973, when still employed by the MRC Laboratory in Cambridge, Brenner published a paper that addressed this very subject. It opened with the lines, 'How genes might specify the complex structures found in higher organisms is a major unsolved problem in biology.' He explained that by now many of the molec-ular mechanisms previously shown in microbes were found to exist in much the same form and function in eukaryotic cells – the nucleated cells of animals and plants. The genetic code was universal and the translation of that code to the mechanisms of protein synthesis appeared to be equally universal. Meanwhile, 'Although there are many theories suggesting how the [DNA of higher organisms might control such] complex genetic regulation, the problem is still opaque.' Brenner chose a new model system for research into how animal genes were controlled and organised. In his paper he introduced his new model, a minuscule round-worm, *Caenorhabditis elegans*, which was just a millimetre long and a common inhabitant of temperate soil environments. *C. elegans* had a number of attractive properties for this type of research. It was non-parasitic, so it wouldn't infect laboratory workers; it was very simple in structure, with the entire worm comprising just 959 cells; it could easily be bred; it was conveniently transparent so one could peer inside it through a microscope; it had a tiny genome comprising just five pairs of autosomes and one pair of sex chro-mosomes; and it comprised two sexes – hermaphrodite and male.

In a nutshell, it presented geneticists with an ideal model experimental animal, being easy to breed, safe to store in large numbers, with a sexuality and a genetics that could easily be manipulated.

In his paper, Brenner showed how he had used experimentally induced mutations in some 300 of the worm's genes to show how these genes contributed to the worm's biological make-up and behaviour. But even in a creature as simple as the worm, the genetics proved to be more complex than Brenner had imagined. A staggering 77 different genes were involved in its simple wriggly movements. Nevertheless, study of the worm soon confirmed his choice of experimental model. Here was a new experimental model capable of figuring out what genes did, and in particular how they regulated the mysterious and profound changes that took place during embryological development, when those extraordinary pluripotent cells of the early embryo began the processes of change that would ultimately give rise to the cells of the many different body tissues and organs.

Brenner's model proved to be an inspired choice. It was taken up in many different scientific centres, and as knowledge grew the *C. elegans*, which itself complemented the earlier fruit fly research, was complemented by pioneering explorations of gene function and gene regulation in fish, frogs, lancelets and mammals in the form of mice – as well as a growing variety of plants.

The human body contains about 200 different types of cell, formed into limbs, tissues and organs, each specialised to perform distinct functions. For the zygote to develop into all of these, it must begin its life as a 'totipotent' cell – a cell that can differentiate into every possible human tissue, including the placenta as well as the developing foetus. The first differentiation is from this stage of totipotent to 'pluripotent' cells – which means cells with multiple but not total potency. The pluripotent cells are the cells that now give rise to the more complex shapes and cellular differentiation that will begin to fashion the distinct tissues and organs. These same pluripotent cells, also referred to as 'stem cells', remain with us for the rest of our lives, replacing damaged tissues in the constant recycling that is necessary for normal physiological functioning and health. For such a remarkable transformation to occur in the embryo with such predictable precision, each cell must know its own fate. This is determined by a carefully controlled bureaucracy of genetic elements including epigenetic regulation, which we shall come to in a subsequent chapter, as well as entities known as 'regulatory genes'.

By the late 1980s geneticists working with the fruit fly discovered a battery of genes that determined the identity of the different segments of the insect body in strict order from the head to the 'tail' end during the embryological formation of the insect body within the egg. They called these 'homeobox' or '*Hox*' genes (the name of a gene is conventionally italicised whereas the name of its transcribed protein is written in ordinary script). Subsequent research would discover that these same *Hox* genes in similar order along a specific chromosome were critically important to the development of the animal body plan during embryological development. We humans, like all vertebrate animals, follow a *Hox*-determined developmental plan that creates a right and a left side. This gives us our 'bilateral' symmetry. We might compare this, for example, to the exotically beautiful sea creatures, called echinoderms, such as starfish and sea urchins, which have a radial pattern of symmetry similar to the segments of an orange or the flowers of a daisy.

As the human embryo grows from the initial ball of cells, our *Hox* genes will determine at which point to place the head, and within the head the eyes, the nose, the jaws; and then, vertebra by vertebra, the seven bones that constitute the neck. Then, vertebra by vertebra again, the twelve bones that designate the thorax, with the offshoots of upper limbs and ribs; in similar fashion, the five lumbar vertebrae that will support the abdomen; and finally the fused vertebrae of sacrum, which supports the pelvis and lower limbs – all in appropriate position along the central axis of our bilateral body plan. The evolution of *Hox* genes was clearly a vital evolutionary step in the evolution of the animal kingdom. So vital is their function that they have been steadfastly conserved by natural selection over vast time periods of evolution. For example, even though the common ancestor of humans and insects inhabited the oceans roughly 600 million years ago, if we were to replace the *Hox* gene for the specification of the insect eye with its human equivalent in the insect zygote, the human gene would still locate the development of the insect eye.

Hox genes code for proteins, but *Hox* proteins are not enzymes, nor do they function in a structural capacity – such as making up skin, or the structures of kidneys, heart or bone. Instead, they regulate the expression, or 'transcription', of other genes. Hence they are dubbed 'transcription factors'. The *Hox*-coded proteins bind themselves to key nucleotide sequences within the chromosomes, known as 'enhancers', where they act by switching the target genes off or on. In time scientists came to discover many such 'regulatory genes' that play important roles during embryological development, as well as in normal human physiology throughout life. Key genes, such as the *Hox* cluster, initiate a process that leads to a series of developmental steps, involving signalling hormones and transcription factors. Key to understanding such systems is the fact that one key gene may activate many secondary genes, each further activating multiple more, so that it ends up with a cascade of hundreds of genes that constitute what is called a 'developmental pathway'. This ensures that a region in the embryo becomes the brain, or a limb, a kidney or a toenail. In fact, if we consider the construction of a complex tissue, such as a kidney or a limb, it is evident that this will contain many different kinds of cells and tissues. For example, a developing leg will include the construction of skin, muscle, bone, nerves and blood vessels – so its embryological development must involve the coordination of many different regulatory pathways, perhaps with local signalling between tissues. A failure of any individual component is likely to lead to catastrophe. Thalidomide was a popular over-the-counter drug used to treat nausea in pregnancy in the 1950s and early 1960s. Within a few years of widespread use approximately 10,000 children were born with severely malformed limbs – so-called 'phocomelia'. The failure of the normal blood vessel development within the developing limb buds was the root cause of the thalidomide tragedy.

At the time of publication of Brenner's paper, in the early 1970s, we understood little of this genetic regulation of human development. We knew, of course, that the human brain was still relatively undeveloped at birth, continuing to grow and develop for perhaps two or three years into infant life. And while we were aware of the glandular changes that brought about and resulted from puberty, we had little or no understanding of its genetic regulation. Today we know that puberty involves very profound changes, both at genetic and epigenetic level – in effect, we return to the controlled turmoil of embryogenesis, in what geneticists now recognise to be a major and dramatic phase of 'postembryonic development'. There are so many similarities in the genetic regulation and the profound bodily changes between human puberty and the astonishing metamorphoses we see in the lives of moths and butterflies, some scientists regard puberty as a variety of metamorphosis.

Prepubertal boys and girls have much the same proportion of muscle mass, skeletal mass and body fat. But now, with the reawakening of the same powerful developmental epigenetic and genetic pathways that controlled embryological development, the bodies of boys and girls undergo dramatic physical change, including rapid growth and major changes in muscle and fat distribution that differ between the sexes. By the end of puberty, men have 1.5 times the skeletal and muscle mass of women, whereas women now possess twice as much body fat as men. These obvious physical changes are accompanied by cellular and tissue changes involving sexual organs and related organs, such as the breasts in women and the prostate glands in men. Puberty is brought about by signalling pulses of gonadotrophin-releasing hormone (GnRH), which is released by the hypothalamic portion of the brain. This stimulates the master gland, the pituitary, to increase the secretion and release of the sex gland stimulating hormones, or gonadotrophins, which travel through the blood stream to the ovaries and testes, where they provoke increasing blood levels of oestrogens and androgens. If sometimes the pubertal adolescent appears moody and confused, it is hardly surprising given the massive physical and hormonal changes that are coming about. Only recently have we recognised that puberty is also associated with hormone-driven major rewiring of the neural circuits of the adolescent brain, triggering changes towards adult behaviour.

Some psychologists have proposed that individual differences in adult behaviour and the risk of sex-related psychopathies may derive from variations in timing and in the interactions between the hormones of puberty and the brain rewiring at this critical period of adolescence.

*

By the early 1990s biologists had a basic understanding of how genes worked. They knew that genes coded for several different types of protein. Some played key roles in our internal chemistry, as enzymes, while others were structural components in cell membranes, the tissues of skin, eyes, hair and nails. Geneticists had plotted where hundreds of genes were to be found on the 46 human chromosomes. They were accumulating substantial knowledge about genetic regulation. They were also beginning to realise that there were additional systems of regulation that were not DNA-determined. There was growing evidence of non-DNA-based systems that could influence the expression of DNA from outside the genome – systems that might be capable of changing during the life and experience of the individual. These would in time come to be understood as part of the epigenetic regulatory system, which will be described in a later chapter.

The revolution that had begun in 1953 with the discovery of the structure of DNA, had given rise to the novel discipline of molecular biology, with its manifold extrapolations to medicine and biology. In a few decades, we had discovered more of the labyrinthine inner workings of human heredity, embryological development and the workings of cells, tissues and organs at biochemical level, than in all of previous history. There was also growing evidence that viruses had invaded our human genome, from the presence of viral genetic sequences and even whole viral genomes. While some geneticists regarded these as junk, the fossils of infection from long ago, others believed they might be contributing in some functional ways.

Thousands of genes had been discovered using the laborious techniques of mutation in experimental animal models. But, given that the human body incorporated an estimated 80,000 to 120,000 proteins, on the basis of one-gene-one-protein there should be as many genes encoding all those proteins. This suggested that there must be vast numbers of unknown genes yet to be discovered. What geneticists now needed to know extended beyond the sequencing of individual genes. The next logical step must be the exploration of the structure of every chromosome, and beyond that the exploration of the entire nuclear genome. Without such exploration we could not determine how the system worked as a whole. Only with the sequencing of the entire human genome would we understand what lay at the core of our being – to paraphrase Bronowski, what genetic gifts made us so unique among the animals. All that we needed to take this final almighty step was the political will to fund the exercise, together with more efficient techniques of reading DNA sequences.

Back in the mid-1970s a Cambridge-based British biochemist, Fred Sanger, already a Nobel Laureate in Chemistry for work on the structure of proteins, had pioneered novel techniques of automated DNA sequencing that were subsequently named after him: 'Sanger sequencing'. Sanger used these techniques to determine the first complete genome of an organism – the same organism I studied as a student, and doctor, the bacteriophage virus known as ΦX174. This won him a second Nobel Prize in Chemistry, the only Nobel Laureate ever to win two prizes in Chemistry. While Sanger's methodology became the standard technique for DNA sequencing in laboratories throughout the world, discovering the structure of tens of thousands of genes, it was, in Sanger's own admission, slow and laborious to conduct, requiring the scientists to read off the results on printouts, and demanding wasteful quantities of radio-active phosphorus which was used to label the nucleotides. In the mid-1980s Leroy Hood and colleagues at Caltech, in America, introduced a faster and easier methodology which labelled the nucleotides with four different-coloured fluorescent dyes that could be read by a laser within a machine. Others developed techniques of replicating genetic sequences using cultures of *E. coli* bacteria, so that small amounts of DNA could be amplified to make sequencing easier. The genome could now be broken down into smaller sequences which could be amplified to many copies in bacteria, and then sequenced in automated machines.

In 1984, the political component achieved critical mass when the United States Department of Energy proposed that they would fund the sequencing of the entire human genome, with its 6.6 billion nucleotide sequences. The name of the project was decided by a committee: it would be 'The Human Genome Project'.

The scope of the project was daunting, but it was also fantastically ambitious, inspiring and exciting. By 1987 the proposal had been debated and clarified, with a crystal clear statement of purpose: 'The ultimate goal of this initiative is to understand the human genome, knowledge of which is as necessary to the continuing progress of medicine and other health sciences as knowledge of human anatomy has been for the present state of medicine.'

Exclusively American to start with, the project later expanded to include many other countries, ultimately growing into the largest collaborative biological project in scientific history. With so many different scientists and scientific groups involved, it was inevitable that there would be conflicting opinions on how to go about it. Some thought we should undertake a chromosome at a time, but this would stretch to maybe ten or even fifteen years to complete. Some politicians failed to grasp the importance of the project and bridled at the likely financial cost, which would rise to billions of dollars. Some people may have been somewhat daunted by the prospect of such a monumental step into the unknown.

But by the early 1990s the die was cast. In 1990, two major funding agencies, the DOE and the National Institutes of Health (or NIH), coordinated their plans. That same year James Dewey Watson, joint discoverer of the structure of DNA, was appointed head of the NIH programme. Watson's prestige, the backing of the US National Academy of Sciences, the support of many academically influential molecular biologists, and the governmental and other official sponsorship of something like $2.6 billion, now underpinned the project. Watson immediately encouraged the internationalisation of their plans, enlisting the help of the UK, Germany and France, with contributions from many other European centres, as well as Japan, China and Australia. In the UK, the Wellcome Trust became a major charity partner with the US public bodies.

With the major opus now organised, coordinated, funded and ready to roll, the banks of computers and the automated sequencing machines kicked into action. It was still the general assumption that the project would take fifteen years to complete, but that would change with the intervention of an unexpected rival from the business sector – an American commercially sponsored organisation named 'Celera Genomics'. The competition from a rival, privately sponsored organisation would throw an unwelcome spanner into some of the most conservatively organised plans.

eight
First Draft of the Human Genome

*I feel this is an historic moment. This is the most important
scientific effort that humankind has ever mounted . . . It will
change biology for all time.*

FRANCIS COLLINS

On Sunday 12 February 2001, the two rival organisations, Celera
Genomics and the publicly funded Human Genome Project –
embodying many different governmental and major charitable
bodies in the US, UK, Germany, Japan and France – announced
simultaneously that they had completed the first comprehensive
analysis of the human genome. It created a wave of excitement
throughout the world's media. In the United States, President
Clinton led the chorus that would be joined by Prime Minister
Tony Blair in the UK, and by similar national leaders and leading
scientific figures in every country, proclaiming that a new epoch
of knowledge and scientific exploration had arrived. In the UK,
Roger Highfield, science editor of *The Daily Telegraph*, put it
bluntly: 'Science Rivals Open the Book of Life'. In the words of

Andy Coghlan and Michael Le Page, writing for *New Scientist*, the genome would soon be as familiar to schoolchildren as the periodic table of the elements. There could be no doubting that it signalled a new milestone in genetic science, the most powerful and logical development to follow the breakthroughs on DNA. And like the DNA story, there were new issues of rivalry and conflict between the discovering groups.

As director of the Human Genome Project, Watson had internationalised the scope of the undertaking, meanwhile engendering the gratitude and dedicated support of academic scientists throughout the world. He had also devoted a small but significant allocation of funds to include sociological, religious and ethical concerns in the minds of lay intellectuals and politicians. Celera Genomics was seen by some within scientific academia as a brash interloper, led by an entrepreneurial scientist, J. Craig Venter – but to his credit Venter had, through insight and sheer force of personality, already succeeded in a litany of impressive breakthroughs involving new fields of genetic discovery. Like Watson, Crick and Wilkins, Venter would also admit that he had been inspired by Schrödinger's book.

Growing up as part of academia, Venter had worked for the US National Institutes of Health in a lab immediately below that of Marshall Nirenberg, who had contributed to the discovery of the histone code. In 1992, impatient with the slow progress of academia, Venter set up his own commercial laboratory, The Institute for Genomic Research, or TIGR. He was now free to combine automated sequencing with his group's newly invented 'shotgun' approach, in which incredibly lengthy genetic sequences found in living genomes could be broken down into smaller fragments. By breaking down the same genome again and again, and breaking the sequences in different regions, his team could use the inevitable overlapping sections to piece together the fragments so that the entire sequence of, say, a microbe, or a human chromosome and so on, to the entire genome, could eventually be stitched together.

This 'shotgun sequencing' technique had the potential to speed things up but it was derided as potentially inaccurate by Venter's academic rivals. Nevertheless, by 1995, Venter was ready to publish his first success – the first ever completely sequenced genome of a bacterium, *Haemophilus influenzae*, which causes respiratory and other infections. This was followed by the genome of the ulcer-causing bug, *Helicobacter pylori*, and in March 2000, the first insect genomic sequence – that of Thomas Hunt Morgan's famous experimental subject, the fruit fly. The sceptical world of academia was rocked back onto its metaphorical heels.

In May 1998 Venter had teamed up with Perkin Elmer to amalgamate his Institute for Genomic Research with Elmer's PE Corporation, and formed a new company, Celera Genomics. Celera, as its Latin derivative implies, conveyed the objective of 'speed'. Its remit, as Venter made clear, was not biotechnology *per se* but the provision of information. In the words of James Shreeve, who wrote the history of this extraordinary period, Celera's market product would be a massive database of DNA, with the human genome sequence as its heart. Thus Venter, and the new company, had as their very *raison d'être* a vested interest in rivalling the publicly sponsored Human Genome Project.

In 1992, James Watson had a major disagreement with Bernadine Healy, then director of the National Institutes of Health, who had been put in charge of the Human Genome Project. Healy supported the dictates of Congress that NIH discoveries should, where possible, be supported by patents. Watson heatedly disagreed. Watson derided Healy until, 'tired of his insulting remarks', she fired him. That same year Watson was replaced as director of the Human Genome Project by the more diplomatically savvy

Francis Collins. In the UK, the Wellcome Trust had set the ball rolling by funding a major sequencing laboratory near Cambridge, the Sanger Centre, which would work in a coordinated synchronicity with NIH on the Human Genome Project.

The ambitious Celera commissioned 200 of the fastest automated sequencing machines, which would combine the speed of mass production with Venter's shotgun method, blasting the 46 chromosomes (containing some 6.4 billion nucleotides) into much smaller fragments which would be capable of decipherment by the banks of the sequencers before being reassembled to complete the whole genome. The Celera approach, as Venter now planned it, would reduce the time needed to complete the project from the ten years proposed by his rivals to seven. Meanwhile Collins, on behalf of the many scientists involved in the publicly funded Human Genome Project, argued that this technique would lead to unacceptable inaccuracy. The academics now raised new worries – that, despite Venter's reassurances, the commercial mindset would lead to unacceptable limitations in the freedom of access to the genomic data, hampering future research. Taken to extremes, some scientists feared that Celera might attempt to copyright our human genes.

This acrimony and debate still rankled between the rivals and infused the media, at the time of the twin declarations of discovery, in 2001, with the Celera results published in the American flagship magazine, *Science*, and the Genome Project's results published in the British equivalent, *Nature*. In effect we now had two readout versions of the same genome. While Celera made clear that they would permit free access to their findings to academic scientists, this would not extrapolate to commercial applications and potential. After all, they had spent hundreds of millions of dollars in the exploration and, as a commercial company, they needed to recoup their costs and make some profit out of the enterprise. Meanwhile, the publicly funded group made explicit that all of their findings were now universally available.

Some readers might feel indignant that commercial interests should intrude into something as sacred as our genetic make-up, but in fact this parrying between commercial and public interest is commonplace in biological and medical research. While it may at times be tricky to draw any hard and fast line between the two very different approaches, in practice research into the most important of arenas, such as vaccination, antibiotics and the treatment of cancer, has always involved an uneasy balance between opposing interests.

The breakthrough actually came about through both avenues of research, and with equal plaudits to the two opposing sides. Through the twin publications of *Nature* and *Science*, the world of science, and humanity in general, was now privileged to learn, on 15 and 16 February respectively, about the enormously complex molecular structures that lie at the genetic heart of us. The decipherment was epochal in what it promised future generations of biological and medical scientists – indeed, in what it promised human society – but it also proved to be mind-blowing in its unexpected revelations. If, as newspapers and magazines proclaimed, here was the basic genetic landscape at the core of life, that landscape was now revealed to be a vast *terra incognita*.

The word 'breakthrough' is often misused in relation to scientific discovery, but here, indeed, was real breakthrough after breakthrough. And the breakthroughs presented a very much unprepared world of science with not one but three major surprises, each a challenging new mystery. These will become evident if we examine the pie chart of the 2001 human genome below.

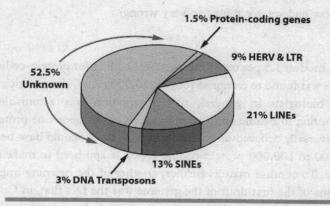

I should make clear that this pie chart is a metaphor of sorts, summing up the percentage contributions of different distinct genetic elements to the genome without reference to where things are actually situated throughout the 46 chromosomes. At this stage in our knowledge most geneticists were mainly interested in genes that coded for proteins, so it is in this aspect, the part of the genome that codes for proteins, where we discover the first of the three mysteries.

Biochemists had arrived at a rough assessment that there were something like 100,000 proteins involved in the structure and functioning of the human body. Thus we anticipated that there would be the same number of protein-coding genes. What many geneticists wanted to know was exactly how many protein-coding genes there really were, and where they were situated on the chromosomes. It was an almighty shock to discover that these protein-coding genes amounted to less than 2 per cent of the entire genome, perhaps as little as 1.5 per cent. It hardly seemed possible that this minuscule inheritance could possibly translate to the 100,000 different proteins that went into the make-up of the human body.

How had we got things so very wrong?

*

This modest 1.5 per cent of the genome coding for protein-coding genes was found to comprise roughly 20,500 genes. For geneticists, and biologists in general, this observation was astonishing. According to Beadle and Tatum's maxim of one gene one protein – universally believed up to this moment – there should have been 80,000 to 100,000 protein-coding genes. It appeared to make no sense. To confuse matters further, another of the dawning implications of the first draft of the genome was the fact that, in Craig Venter's estimation, at least 40 per cent of the protein-coding genes they had discovered had no known function. 'We have no idea what they do. They have not been seen in biology before.' He went on to admit: 'It is incredibly humbling.'

The paltry 20,500 genes seemed downright humiliating. To put it into perspective, we had roughly ten times as many genes as the average bacterium, four times as many as a fruit fly, and just twice as many as a nematode worm. In terms of genes, we seemed hardly more complex than these humble life forms. A related revelation was the number of genes we have in common with these simpler organisms. We now discovered that we share 2,758 of our genes with the fruit fly and 2,031 with the nematode worm; and the three of us — human, fly and worm — have 1,523 genes in common.

Darwin had been the first to dare to imagine that all of life on Earth was intimately related, through the process of evolution that he had himself pioneered. Here at once was both the confirmation of his brilliance in the letters of the code of life, our human DNA, but also a new and astonishing incongruity.

How could science possibly explain how roughly 20,500 genes could code for the estimated 100,000 proteins?

Up to this point we believed that the protein-coding genes, made up of long strands of DNA, were copied to their exact matches in terms of complementary messenger RNA – with the exception that the fourth nucleic acid, thymine in the DNA, was replaced by uracil in the RNA – and this matching long strand of messenger RNA was then ferried out of the nucleus and taken to the protein-manufacturing ribosomes in the cytoplasm, where it was translated, using the triplet codes, to proteins whose amino acids corresponded faithfully to the original DNA code of the gene in the nucleus. Thus the number of genes should correspond to the number of proteins.

The key to this enigma proved to be a startling discovery, made by two scientists back in 1977.

Richard J. Roberts graduated from my own alma mater, the University of Sheffield, with a Bachelor of Science degree in chemistry, completing his PhD in 1965. He subsequently went to work at Cold Spring Harbor Laboratory, New York. Phillip Allen Sharp graduated at the University of Illinois with a PhD in chemistry in 1969. He also ended up working at Cold Spring Harbor. Roberts and Sharp were exploring how the genes of a virus, called adenovirus 2, were expressed as protein within the cells of tissue cultures. What they discovered was that the actual messenger RNA strand that arrived at the ribosomes ready to code for the protein was significantly shorter in terms of its nucleotide sequence than the DNA-based gene in the viral core. This told them that only a portion of the so-called protein-coding gene actually coded for the amino acid sequences of the translated protein. Something very strange must have taken place during the chain of communication from the viral gene, within the viral core, and the expression of that gene within the host cell in the tissue culture.

As with the phage research a generation earlier, the tiniest of microbes, the viruses, had opened up a window onto a more general biological truth. Roberts and Sharp had discovered what we now call 'introns' and 'exons' and the importance of their role in a genetic mechanism known as 'splicing' – discoveries that led to their sharing the Nobel Prize in Physiology or Medicine in 1993.

What then are introns and exons? And how do they solve the puzzle of the discordance between the number of protein-coding genes and the anticipated number of proteins coded by the human genome?

Perhaps it is time we climbed back aboard our imaginary train to take a new journey into that ultramicroscopic landscape, with its astonishing twin track of alternating phosphates and deoxyribose sugars, and those all-important sleepers.

*

We arrive at our destination in the blink of an eye to find ourselves chugging along large stretches of a chromosome. We know that within this chromosome there are distinct stretches of DNA called genes. Since this is a wonderland, with magical potential, we can wish that some forthcoming gene should show itself by glowing with a green light. With this in mind we slow down sufficiently to see that exactly such a stretch is looming in front us, pulsating a beautiful emerald green, which tells us that we have arrived at the beginning of a gene. We throw the engine into low gear and travel along the twin-track rails, observing that the green glow is actually coming from the sleepers. After a while we see that the track has reverted to the normal brown of sleepers again. I must now suggest that we haven't actually come to the end of the gene. The green-glowing track we have just traversed is merely the first exon.

You are inclined to ask: 'So where exactly are we now?'

'The normal stretch, with the brown sleepers, is the first intron.'

As we chug slowly along this section, we find it is, if anything, longer than the green-glowing previous section. Then it too ends abruptly, as we arrive at another green-glowing section – a second exon. As we continue our journey, we count some three stretches of exons with two intervening stretches of introns. There are no further green-glowing sections. So what we have been looking at is a protein-coding portion of a gene comprising three separate exons with two introns, somewhat like spacers, in between them. It really is that simple. What Roberts and Sharpe discovered is that the whole DNA of a single 'gene' does not necessarily code for a single protein. The gene is actually broken down into smaller chunks, the exons, separated by intervening introns. To code for a specific protein only a particular cluster of the gene's exons will be expressed – they will be copied to messenger RNA, complete with the intervening introns, but the intervening introns will be removed from the coding sequence before the exons are then 'spliced' together to fashion the final messenger RNA that will code for a protein.

It might help us remember if we think of the exons as 'exiting' the nucleus to make proteins, while the introns stay 'in' the nucleus. The total number of exons in any one human gene is very variable, with an average of 8.4. So in order to make a specific protein the genome must know how to pick out the right gene, and then, within the gene, must be capable of choosing which exons to splice together to code for the relevant protein.

Take, for example, our human beta-globin, which is part of the molecule haemoglobin. We now know that haemoglobin contains a single iron atom at its core, surrounded by two alpha protein subunits and two beta protein subunits. So the protein as a whole is made up of four different parts – it's a so-called quaternary protein. Now if we look at one of the two identical beta-haemoglobin subunits, the same protein subunit that is mutated in sickle cell disease, we find that the DNA that codes for these comprises three exons with two intervening introns.

It might help at this stage to know how a gene is activated.

If we were to alight from our train and take a look at the actual stretch of DNA that codes for beta-haemoglobin, we would find that, somewhere close to the start of the first exon (remember that the decoding mechanism moves from the left and moves along the DNA molecule to the right) we find a section of DNA known as the 'promoter'. Somewhere more distant – maybe at some considerable distance – there are other stretches of DNA that act as 'up-stream regulatory elements' – another office or maybe several offices full of administrative bureaucrats. The bureaucrats send a signalling wire to the promoter to say, 'Time to express the gene.' Whether or not a specific gene is expressed will vary from cell to cell, tissue to tissue, organ to organ, within the human body, and so will the timing of gene expression and the amount of gene expressed. That's what the bureaucrats control. The promoter then instructs the gene to express its DNA. In the case of the beta-globin protein, the three exons, together with the two intervening introns, are converted to the matching messenger RNA, after which, and still within the nucleus, the two introns are excised and the remaining three exons are joined up together. Only now does the messenger RNA leave the nucleus and travel to the protein-manufacturing ribosomes in the cytoplasm.

The largest known human gene codes for a protein called 'dystrophin', which has 79 exons separated from one another by 78 introns. Dystrophin is important for normal muscle function. As with sickle cell disease, mutations affecting this very long protein can give rise to an inherited form of disease. For example, in Becker and Duchenne muscular dystrophies, a whole exon is usually missing. This damages the membrane surrounding the muscle fibre, resulting in impaired muscle function.

Understanding of the genetics of diseases like these can help medical scientists to work on a treatment and, perhaps in the not too distant future, work towards a genetic cure. Moreover, understanding of how exons and introns work now affords an explanation of how just 20,500 genes could possibly code for 80,000 to 100,000 proteins.

A gene which, for example, had 14 exons separated by 13 introns, is likely to code for more than one protein. All that is necessary is that the regulatory mechanisms, which decide on which exons to splice together to make the messenger RNA, choose different combinations of exons. We now know that this is exactly what happens. The ability of a single gene to code for more than one protein is known as 'alternative splicing'. We also know that this is ubiquitous in eukaryotic life – including all the animals, plants, fungi and simpler forms whose genome is contained within a nucleus.

Now we understand why the Nobel authorities decided to award Roberts and Sharp with the Nobel Prize in Physiology or Medicine in 1993. In 2005 a multi-million-pound expansion to the chemistry department of the University of Sheffield, where I once studied, was named after Richard J. Roberts.

*

As we have seen, the first of the major enigmas thrown up by the 2001 blueprint of the human genome had a ready solution. But the other two, the vast virus-related segments, and the empty 50 per cent, will take a good deal more explaining. Before we journey into these more difficult territories, we require a basic understanding of the mechanisms that are capable of changing the genomes of existing species, and in doing so, of creating new life forms. This will require a basic understanding of evolutionary biology together with some very recent discoveries within this broad, exciting discipline.

nine

How Heredity Changes

. . . in a dozen years, The Origin of Species *has worked
as complete a revolution in biological science as the* Principia
*did in astronomy – and it has done so, because, in the words
of Helmholtz, it contains 'an essentially new creative thought'.*

THOMAS HENRY HUXLEY

When, in 1859, Darwin first published his theory of evolution in
his book, *The Origin of Species by Means of Natural Selection*, it
provoked a tsunami of shock throughout the civilised world.
Although he made little or no reference to human evolution in
this, his first book, the implications for human evolution were
implicit in every thought and line. Given that there was no real
understanding of how heredity worked, his thinking remains
remarkably prescient today. In essence he proposed that nature
selects for key characters, or 'traits', that enhance the potential
for survival in the same way that breeders of domestic animals
and crops had long selected for traits such as size of kernel, coat
of wool, meatiness of muscle, resistance to disease, or drought,
and so on. The way nature did so was brutal, though: it was
through attrition. Most parents, for example in animals or plants,

produced far more than two offspring. Yet by and large the numbers within a species stayed roughly constant. Darwin realised that the offspring had to compete with one another for scarce resources or to avoid predators. This created fierce competition for survival; those who had a slight edge in the tooth and claw of nature were more likely to survive. If this edge was determined by heredity, the survivors would pass it on to their offspring. In time – and Darwin was well aware that this would most likely involve a gradual and incremental sum of small advantages over very long periods of time – the advantaged would be more likely both to multiply and eventually to generate descendants sufficiently different from the original parental strain as to give rise to a new species. Dilution of the hereditary advantage would be reduced if the emerging species was geographically isolated from the parental strain – for example through isolation on islands, or through separation by mountains or major rivers. In time the new species would be sufficiently different physically, and reproductively, to breed true within its own population.

Natural selection was a very simple and convincing hypothesis. Darwin had observed differences in the beaks of the birds on the different Galapagos islands. Soon other naturalists – what today we call biologists – would observe and confirm his findings in animals and plants, fungi, protists (what we once called protozoa) and much simpler organisms, such as bacteria and viruses.

While many scientists were intrigued by and largely supportive of Darwin's theory, some, such as the distinguished Swiss-American Jean Louis Rodolphe Agassiz, who had performed landmark work on glaciers and extinct fishes, were adamantly opposed to any theory of evolution on religious grounds. Darwin's former friend Sir Richard Owen, the renowned naturalist and founder of the Natural History Museum in London, is also presented as opposing evolutionary theory on religious grounds, but it would appear that he had his own theories about it and simply disagreed with Darwin's concept of natural selection combined with gradual change. Darwin was well aware that natural selection could only work if there were mechanisms capable of creating changes in the heredity of living organisms. To put it another way, natural selection requires hereditary variation for it to work. Some of the resistance from within science itself derived from the prevailing lack of understanding about the nature of heredity. In the subsequent opinion of Sir Julian Huxley, grandson of Thomas Henry Huxley, who championed Darwin in his lifetime, it was this lack of understanding in particular that dogged confidence in Darwinian theory as science moved towards the end of the nineteenth century. In the opening chapters of his book, *Evolution: The Modern Synthesis*, Julian Huxley put his finger on the heart of the problem: 'The really important criticisms have fallen upon Natural Selection as an evolutionary principle and centred round the nature of inheritable variation.'

It was hardly a criticism of Darwin that he could not explain how hereditary change might come about – next to nothing was known about it in his day. He speculated that hereditary variation arose from a kind of 'blending' of the pedigrees of the two parents. The first two chapters of *The Origin* are devoted to explaining how blending worked, both in animals and domesticated crops. But over time, Darwin himself became less and less convinced that blending was a sufficient explanation. In the words of the leading American Darwinian, the late Ernst Mayr: 'The origin of this variation puzzled him all of his life.' Today we know that what Darwin implied by 'variation' suggests some mechanism or mechanisms that give rise to hereditary genetic, or genomic, change. The rediscovery of Mendel's laws of heredity produced a breakthrough in the understanding of how heredity actually operated: specific characters, or traits, were inherited as discrete genetic units – what we now call 'genes'. In 1900 a Dutch botanist, Hugo de Vries, took this an important step further when he had the inspiration that heredity could be altered if mistakes were made during the copying of these genes. The obvious opportunity was during reproduction – here a mistake in copying a gene would give rise to what de Vries called a 'mutation'.

In the 1920s and 1930s the reality of mutations was confirmed by the laboratory experiments of evolutionary biologists such as Thomas Hunt Morgan, Barbara McClintock and Hermann J. Muller. Mutation was no longer a theoretical possibility but a fact and, very likely, a common enough fact to be mathematically predictable. A number of investigating scientists throughout the world began to piece together a mathematically based synthesis of how natural selection would be enabled through such 'germ-line mutations' of genes. These included pioneering geneticists such as Ronald Aylmer Fisher and John Burdon Sanderson Haldane in Britain, Sewall Wright and Theodosius Dobzhansky in the United States and Sergei Sergeevich Chetverikov in the Soviet Union.

In time geneticists found that most of the mutations in DNA sequences during the formation of the human ova and sperm had little or no effect on the function of proteins, and thus seemed unlikely to contribute to evolution or to disease. Those that did give rise to change in proteins, or regulatory function, mostly did so for the worse. They were the causes of inherited diseases. But a small minority of mutations altered the heredity of the offspring in ways that might potentially improve the chances of survival. For example, there is growing evidence that a small number of mutations in a gene known as *Prx1* may have contributed to the elongation of the forelimb skeleton that enabled the evolution of the membranous wings of bats.

From a medical perspective, mutations of DNA can also arise during the cell division that is a normal part of the replenishment processes in the many tissues and organs during life. These so-called 'somatic mutations' are important in the causation of various types of cancers, from the leukaemias and lymphomas of blood and lymphatic tissues to cancers of breast, skin, kidney and bowel, and so on. The reality is a little more complex. The genomes of the eukaryotic life forms – those with nucleated cells, including animals and plants – have mechanisms for correcting these copying errors as they arise, but these mechanisms can sometimes fail or be overwhelmed.

Medical geneticists can now list thousands of germ-line mutations that give rise to a range of inherited problems affecting the internal chemistry of the affected offspring. Many of these 'errors of metabolism' arise from a mutation affecting a single gene, but some can result from mutations affecting clusters of genes, aberrant sections of chromosomes or the loss or gain of a whole chromosome. In an earlier chapter we witnessed the recessive mutation affecting beta-globin that causes sickle cell disease. At this stage we might hop aboard our magical train to visit the genome of an individual who has had the misfortune to inherit a dominant mutation, and take a look in a little more detail at how this mutation has come about.

Each of our 46 human chromosomes is, in our model, a separate railway line. Trains can only run from start to finish – they cannot switch lines, since each chromosome is a separate linear structure. On this occasion we choose to travel on Line 4 – human chromosome 4 . We chug along until we come to a stretch of track that is signposted 'Huntingtin'. If we alight here and examine the adjacent track carefully, we observe the typical gene structure we saw in an earlier trip. Here is the section of DNA, with its nucleotide sleepers, that announces itself as the Huntingtin gene 'promoter'. This sequence, which is usually adjacent to the start of the gene, is the genetic switch that turns the gene off and on. From here we stroll further eastwards, moving 'sense-wise' along the track, till we arrive at the first exon of the gene. As we stroll a little further along the exon track, we come across something very odd; we see that a triplet of nucleotide sequences, cytosine-adenine-guanine – CAG – appears to be repeating itself over and over in successive sleepers.

'Go ahead,' I suggest. 'Count the number of repeats.'

You are surprised to discover that there are 45 CAG repeats, one after another, in the first exon of the gene, Huntingtin.

'This mutation is the cause of the illness called Huntington's disease, which causes cerebral deterioration during adult life.'

'You mean there should be no repeats?'

'It's a little more complex. Curiously, we all have many repeats of the CAG sequence in the first exon of the gene, Huntingtin. It's the actual number that determines whether or not we inherit the condition. If we have between 6 and 34 repeats, we do not inherit the disease. The more repeats above this number the more likely we are to inherit the condition. Above 40 repeats means disease in nearly every case. And the higher the number the younger the onset of symptoms.'

'So what we find here is bad news for this unfortunate individual?'

'I'm afraid it is. All humans have two versions of chromosome 4, one inherited from our mother and one from our father. If we were to go and visit the other version of the gene on the matching chromosome, we'd discover that it was normal.'

'In other words, Huntington's disease is, what . . . a dominantly inherited mutation?'

'That's right. It also means that if medical science could find a way of switching off this damaged gene, the remaining normal gene would take over and the condition would, hopefully, be cured.'

At first people only thought of mutations as affecting these types of protein-coding genes. But as geneticists came to understand the importance of genes that coded for regulatory genetic sequences, including genes that coded for proteins that were intrinsically involved in gene regulation, they realised that a mutation that affected a regulatory sequence, for example a sequence that affected embryological development, could also affect the physical and mental development of the offspring. We shall look at this in more detail in later chapters. At this point I merely wish to explain that the same patterns of mutation will sometimes change the hereditary nature of an individual in a beneficial way – a way that enhances the individual's chances of survival. And since this is hereditary, that beneficial mutation will be passed down to the individual's offspring and future generations. This doesn't just apply to humans; it applies to all animals, plants, fungi – indeed to every living organism. This is integral to the way in which evolution operates.

For almost a century, evolutionary geneticists have been recording how mutations in protein-coding and regulatory regions have contributed to the diversity of life on Earth, from the evolution of whales and dolphins from original land-living creatures to the origins of flight in insects and birds. They have also found some evidence for the evolution of genes that may have contributed to the expansion in size, and complexity, of the human brain. But mutations didn't have to be as dramatic as this. A small change that affected, say, the duration of effectiveness of a digestive enzyme such as lactase in humans, is capable of telling us a great deal about our own migratory history. As we shall discover in later chapters, evolutionary genetics appears to be entering a golden age, where the genomes of long-dead ancestors, including supposedly extinct humans, are being resurrected and subjected to intensive study. Soon we shall be in a position to determine with clinical accuracy why people of European origins found a way to digest cow's and goat's milk throughout life while those from Asian ancestry did not.

We are already capable of determining through resurrected genomes when and how Europeans developed blue eyes and fair, or red, hair, in the same way that we can determine, through the genomic examination of fossil bones, how dark-skinned our ancestors were – or through analysis of teeth, how fast they matured during childhood and what diet they consumed.

The inspiration, and subsequent study, of mutation has provided evolutionary biology with a treasure trove of information on how life evolved and diversified on Earth. But the fact that the mutations occur randomly – and this random accumulation of mutations is easily measured – is only part of it. Random mutation on its own would not be enough to create biodiversity. The key to understanding is that natural selection is operating on the variation being presented by the random mutations. And the operation of natural selection is not random; it chooses those mutations that favour survival, and through survival, reproduction.

Mutation plus selection was soon recognised to be a very important mechanism in evolution. It has played a major role in the evolution of the human genome. It also has a seductive mathematical attraction: since mutations are thought to arise at a fairly regular rate – giving rise to what we shall subsequently encounter as the so-called molecular clock – mutation plus selection lent itself to calculus-based mathematical extrapolations that were increasingly seen as the major, if not the exclusive, mechanism of evolutionary change. This came to be viewed as the central mechanism of modern Darwinism, also called neo-Darwinism. Today many school and college teachers still teach that this is the main, if not the only, source of the hereditary change, but we now know that mutation is not the only mechanism of creating hereditary change. On the contrary, mutation is one of a number of different naturally occurring mechanisms that are capable of changing the heredity of living organisms.

For close to a century biologists and molecular geneticists have been gathering information on three other mechanisms that also generate the hereditary change necessary for evolution to take place. These include epigenetic inheritance systems, genetic symbiosis and hybridisation, which, together with mutation, I have gathered under the convenient umbrella designation of 'genomic creativity'. I chose my words carefully when I coined the phrase in a paper published in the *Biological Journal of the Linnean Society* because I wanted to emphasise that these four mechanisms are creative in themselves. And I used the word 'genomic' rather than 'genetic' because the very definition of epigenetic systems defines them as non-genetic. Each of the three other mechanisms is very different from mutation, and their genetic and genomic implications are also quite different. Following publication of the same ideas in my book, *Virolution*, Gordon N. Dutton, Emeritus Professor at Glasgow Caledonian University, suggested I use the easily remembered acronym MESH for these four distinct mechanisms: mutation, epigenetics, symbiosis and hybridisation. Thank you Professor Dutton, henceforth I shall. As we originally saw with mutation, all four mechanisms work hand in glove with Darwin's concept of natural selection.

ten

The Advantage of Living Together

Had there not been a lack of communication between my teachers and colleagues at Berkeley . . . and my quantitative friends at the Bacteria and Virus Laboratories, I might not have found myself groping with the problems whose possible solution is presented in this book.

LYNN MARGULIS

The study of nature has amply confirmed Darwin's insight – the land, air and oceans are replete with examples of the struggle for survival. Competition for resources, the need for camouflage, the armour of protection, the massing of numbers, such as we see in the great herds of herbivores, shoals of fish and the magnificent flocks of birds, are all evolved strategies for survival in a predatory world. From these very obvious adaptations to the microscopic mutations affecting genes, the evolutionary processes are now seen to be universal. In 1976, Richard Dawkins, while based at the University of Oxford, consolidated two decades of evolutionary study in his iconoclastic book, *The Selfish Gene*, which was seen by many scientists as the perfect modern encapsulation of Darwin's original vision. However, while the concept of competition – which is the main driving force in the vision of both Dawkins and Darwin – is commonplace in nature, it is not the only driving factor in the struggle for survival.

In 1878, at a time when Darwin was still alive, a German professor, Anton de Bary, drew attention to the fact that different life forms sometimes gained an advantage through living together. He called such living interactions 'symbioses'. It was hardly a new observation. Herodotus described how the plover was known to take leeches out of the mouths of crocodiles, Aristotle observed a similar relationship between a bivalve mollusc and a crustacean, and Cicero was so impressed by many such examples to draw the moral that humans might learn from such friendships in nature. Honey bees appear to have an intimate relationship with flowering plants, with the plant supplying the bees with nectar, meanwhile the bees assist in the transfer of pollen to other plants, thus enhancing their success in reproduction. In the cleaner stations of the oceans, predators such as sharks and groupers line up as if arriving at a taxi rank, to have parasites and debris cleaned from their skins and mouths by tiny fish or shrimps. Anywhere outside the cleaner stations and the small fish and shrimps would be seen as food.

In the late 1800s de Bary and another German naturalist, Albert Bernhard Frank, put the study of symbiosis onto a more firm scientific footing, defining the concept and pioneering the study of the biological and evolutionary implications. It is a common mistake to think of symbiosis exclusively in terms of mutualism. Let us immediately clarify the fact that symbiosis is not about Mr Nice Guy, who comes along and shakes hands with Ms Nice Girl and everything is hunky-dory from then on. Only one of the partners needs to benefit for the association to be regarded as a symbiosis. In fact, symbiosis often begins with outright parasitism, which may progress to mutualism. Biologists who study symbiosis today see many examples that would be placed somewhere in between the two extremes. Even in its mutualistic form, symbiosis is about tough bargaining and hard compromising, with survival of the partnership, and thus the partners, depending on the outcome.

One of the first such living associations to be studied by naturalists were the lichens that coat rocks and stones, like the monuments of Stonehenge. Lichens had previously been categorised as a formal branch of the biological tree, with a variety of different genéra and species. But now they were shown to be not species at all but intimate partnerships of algae and fungi.

Frank discovered something very important about the association of fungi and plants in general. When folks go to a garden centre to buy some seedling plants in their pots, they have little idea that what they assume are roots when they shake the root ball out of the plastic pot are for the most part a ball of fungus. All of the land plants have fungal partners that grow into their roots to fashion an intimate symbiosis, with the plant supplying the fungus with carbohydrates for energy and the fungus supplying the plant with water and minerals. This arrangement is called a 'mycorrhiza', which literally means a fungal root. Some woods are underpinned by a vast mass of fungi underground that extends as a contiguous living system to feed the entire wood.

There are a few simple terms we need to grasp. The study of symbiosis is called 'symbiology' and the biologists who work in this discipline are called 'symbiologists'; the interacting partners in a symbiosis are called 'symbionts'; and the partnership as a whole is called a 'holobiont'. As we have seen, symbiosis includes the smash-and-grab of parasitism, where only one of the partners benefits at the expense of its partner, as well as mutualism, in which two or more partners share the spoils. Today we know that symbioses are omnipresent in nature, from the coral reefs to the prairies and from the rain forests to the wind-blasted valleys of Antarctica. From its inception, the definition of symbiosis implied that it was a force in evolution, referred to as 'symbiogenesis'. Symbiotic partnerships also include different types of partnerships, depending on what is being shared. The root symbioses of plants involve the sharing of the products of living chemistry, or 'metabolism', of plant and fungus, so these are called 'metabolic symbioses'. Other metabolic symbioses include the partnership of alga and fungus in lichens and the gut bacteria that play an important role in human nutrition and immunology. Meanwhile, the cleaner station symbioses involve a sharing of behaviours, so these are called 'behavioural symbioses'.

Symbioses, as they become established over long periods of time, will inevitably bring about genetic changes in the partners. Take, for example, some 319 species of hummingbirds, which are widely distributed throughout the warmer parts of the Americas – these live almost entirely on nectar, which is provided by flowers. Specialised joints in the wings of hummingbirds enable them to beat so fast they are practically invisible; this 'adaptation' enables them to hover with pinpoint accuracy in front of the appropriate flower. In this symbiotic partnership, the Columnia plant has changed the shape of its flower to suit the elongated and curved bill of the violet sabrewing hummingbird that pollinates it; meanwhile the hummingbird has changed the length and shape of its bill to exactly fit the flower. If one sits back for a moment to think about this, bird and plant are now influencing one another's evolution to accommodate the symbiosis. To put this in evolutionary terms, natural selection is now operating, to a significant degree, at the level of the partnership – the holobiont.

The benefit of such a mutualism is clear. Only the violet

sabrewing bill fits the Columnia flower: only the Columnia flower is likely to be fertilised by the dab of pollen transferred from flower to flower on the brow of the sabrewing hummingbird.

A third type of symbiosis, known as 'genetic symbiosis', is more powerful still as an evolutionary force.

The most abundant element in the atmosphere is gaseous nitrogen, which must be bound up into more complex chemical compounds to be useful for the internal chemical processes of life. The chemical fixation of atmospheric nitrogen is an essential step that makes the free atomic element available to every animal and plant, yet the ability to fix nitrogen is impossible for all animals and plants working by themselves. It is only found in bacteria. Legumes, such as peas and clover, form symbiotic unions with nitrogen-fixing bacteria, known as rhizobia, that live in nodules within their roots. The rhizobia get the high energy they need from their plant host while the host gets nitrogen in a suitable organic form for its internal chemistry from the bacteria.

But there is an additional wrinkle to the nitrogen cycle. Most species of the rhizobial bacteria that live in soil are not capable of fixing nitrogen. They only become capable when a 'symbiotic island' comprising a package of six genes is transferred into their genomes from a nitrogen-fixing species. This transfer of pre-evolved and ready-to-go genes from one species to another is a very different mechanism of hereditary change from what we saw with mutation. It's an example of what is called a 'genetic symbiosis'.

Unlike the accidental nature of mutations, genetic symbiosis adds genes with pre-evolved potential to a different evolutionary lineage. Some biologists will describe this as 'horizontal gene transfer', which indeed it is. But this is a generically collective term rather than a scientifically definitive concept. The concept of genetic symbiosis defines and explains exactly how the transferred gene came about, and how the mechanism of transfer operates. Like mutation, this genetic change is hereditary: the offspring of the changed rhizobial species will inherit the symbiosis island. And again, just as with mutation, the genetic symbiosis will only become evolutionarily significant if and when it becomes incorporated into the evolving species gene pool by natural selection. Genetic symbiosis, working hand in glove with natural selection, has obvious potential for evolutionary change. At its most powerful level, where it involves the union of entire pre-evolved genomes, genetic symbiosis will create a novel 'holobiontic genome', which brings together the pre-evolved interactive genetic potential from two, or more, quite different evolutionary lineages.

*

Between three and two billion years ago the Earth had no sign of the green life of plants we are familiar with today. It was populated by the first cellular life forms, which comprised bacteria and bacteria-like organisms, called archaea. The atmosphere at this time contained no oxygen. But many of the genetic and biochemical pathways now common to life evolved during this microbial stage, so it isn't altogether surprising that all of life today has many genes, and biochemical pathways, in common. Then, about two billion years ago, life underwent two enormous changes that were described by the eminent evolutionary biologist John Maynard Smith as major transitions. A variety of ocean-bound bacterium, known as the cyanobacteria, evolved the ability to capture the energy of sunlight – the process we call photosynthesis. In time those cyanobacteria, and a variety of other photosynthetic microbes, became part of the evolution of the kingdom of plants, where the microbes have evolved to the organelles in the cells of the leaves that we call chloroplasts. As a by-product of photosynthesis the bacteria excreted gaseous oxygen into the oceanic water and ultimately the atmosphere. Today most of the oxygen in the

Earth's atmosphere finds its way through the photosynthesis of plants, algae and the cyanobacteria that still grow with great abundance in just about every terrestrial and aqueous environment. But this proved to be a catastrophe for the sulphur-breathing bacteria and archaea that originally inhabited the surface waters of the oceans, for whom oxygen proved to be a lethal poison. Today the descendants of these sulphur-breathers are forced to eke out an existence in places where oxygen cannot get to them, such as the insides of animal intestines, or deep in anaerobic mud or between the layers of rock miles under the ground.

Perhaps two billion years ago another species of bacteria made the leap to breathing oxygen. And now a second major genetic symbiosis came about, leading to all of the life forms that breathe oxygen today, including plants, animals, fungi and a variety of single-celled organisms.

How do we know about these extraordinary symbiotic events from the very far distant past? We know because the chloroplasts in the green leaves in plants still retain enough of their original microbial structures and genomes to tell us – and because the mitochondria in the cytoplasm of our human tissue cells also retain their bacterial shapes and structures, and the residuum of their original bacterial genomes. We also know that whereas the evolution of chloroplasts happened again and again, involving different photosynthetic microbes, the symbiotic union that led to mitochondria only ever happened once. Or at least only one such union gave rise to the mitochondria that populate the cells of all animals, plants, fungi and the oxygen-breathing protists that are found throughout biodiversity today. My late friend, Lynn Margulis, pioneered our understanding of the symbiotic origins of chloroplasts and mitochondria, through the serial endosymbiosis theory, or 'SET', which she published in a pioneering book on the origins of nucleated cells.

This symbiotic origin of our human mitochondria is important to our understanding of how the two genomes, mitochondrial and nuclear, still function as a 'holobiontic' union even today.

*

At the time of first symbiotic union, the ancestral bacterium would have probably possessed roughly 1,500 to 2,000 genes. Today, as a result of natural selection working at the level of holobiontic union, the genome of the mitochondrion has been whittled down to a residuum of 37 genes. At some stage in the past, approximately 300 of the original bacterial genes were transferred to the nucleus, where many continue to play a part in the nucleus-mitochondrial genetic linkage that is necessary for normal function. Our human mitochondria populate the cytoplasm, the part of the cell outside the nucleus, where they have evolved to sausage-shaped organelles that look exactly like the original bacteria. The mitochondria also reproduce themselves by bacterial-style budding independently of the reproduction of the nucleus.

This changes the inheritance of diseases that come about through mutations affecting the mitochondrial genes. Where the nuclear genome is inherited from both our parents, and follows the typical Mendelian laws of inheritance (including the patterns of recessive, dominant and sex-linked inheritance we saw in an earlier chapter), the mitochondrial genome is inherited exclusively from our mothers and it follows non-Mendelian patterns of inheritance.

Mitochondria fulfil an enormously important cellular function – enabling our living cells to breathe oxygen. This is further linked to multiple cellular functions, including energy production, the generation of toxic free radicals that are by-products of respiration, and the regulation of programmed cell death, or apoptosis, which is a necessary part of the cycling of cells in tissues and organs. Since the mitochondrial genome is much smaller than the nuclear genome – some 16,500 nucleotide pairs compared to 6.4 billion nucleotide pairs – we might anticipate fewer mutations and thus a low prevalence of genetically induced disease. However, where most of our nuclear DNA does not code for functional proteins, so that mutations are less likely to cause disease, nearly all of our mitochondrial DNA is coding and thus mutations are much more likely to cause disease. Moreover, because it comprises bacterial genes, which are more error-prone than vertebrate genes, mutations in mitochondrial genes are about ten to twenty times more common than would be expected. This is further complicated by the fact that mitochondrial disease can also result from mutations affecting those 300 genes that crossed over into the nucleus. All of this means that we are particularly intolerant of mitochondrial mutations, which are apt to cause serious difficulties with the oxygenation of our living cells.

Mitochondrial diseases are complex and tend to be highly specific to the individual, or family, ranging in severity from mild to fatal. It is hardly surprising that the complexity of the underlying genetics, coupled with the variation in disease presentation, can make the genetic basis of such diseases hard to diagnose and trace. Roughly 1 in 7,600 births are affected by genetic abnormalities affecting the mitochondria, contributing a significant proportion of inborn errors of metabolism in newborn children. Mutations, leading to significant disease, have been identified in more than 30 of the 37 mitochondrial genes and in more than 30 of the related nuclear genes. The illnesses include 'Complex I deficiency', which accounts for roughly a third of all 'respiratory chain deficiencies'. Often presenting at birth or in early childhood, affected individuals suffer a progressive degenerative disorder of the brain and nervous system, accompanied by a variety of symptoms in organs and tissues that require high energy levels, such as brain, heart, liver and skeletal muscle. Another mitochondrial syndrome, presenting in adult life, is Leber's hereditary optic neuropathy, which is one of the commonest inherited forms of eye disease. Most cases of Leber's syndrome are caused by mutations in mitochondrial genes.

There is growing evidence that mitochondrial dysfunction plays a significant role in a much broader spectrum of diseases, and perhaps even the ageing process. Given the advances in genetics, we may in time develop effective gene-based therapy for some of these conditions, but any such therapeutic approach will need to consider the symbiotic evolutionary origins of mitochondria and the complex genetic and molecular dynamics that arise from such an inheritance.

There is another microbe that is quintessentially adapted, through the nature of its life cycle, to entering into holobiontic genetic unions with the genomes of its hosts: this is the rather strange, and in my view, rather extraordinary, microbe we know as a retrovirus.

eleven
The Viruses That Are Part of Us

When I am asked whether poliovirus is a non-living or a living entity, my answer is yes.

ECKARD WIMMER

Eckard Wimmer is a distinguished German-born virologist who has spent his professional life working in America. In 2002 he astonished the world when he and his group of colleagues reconstructed the polio virus from mail-order components they reconstructed in the laboratory. Twenty years earlier, Wimmer had been the first to sequence the polio virus genome. Even today, as his definition suggests, it is head-scratchingly difficult to define what we mean by a virus. This definition has not become any easier with the passing of the years, a difficulty compounded by the recent discovery of giant viruses with 1,000 or more genes, making them genomically more complex than small bacteria. Perhaps rather than attempting to define viruses a more sensible approach is to examine some of their basic properties.

All viruses are coded by genomes – just as in all of life, from bacteria to mammals. Most viral genomes are DNA based, but some have genomes based on RNA. In fact, viruses are the only organisms that use an RNA code. This makes some biologists wonder if RNA viruses might date back to a purported stage in evolution known as the RNA world, which, if this theory is correct, would have preceded the present DNA-based world. RNA, unlike DNA, is capable of replicating without the help of protein enzymes. Thus it would have entailed a smaller step in the origins of life from the purported ambient soup of chemicals for RNA-based self-replicators to set the ball rolling. Viruses are obligate parasites; they are invariably born within the cells of their hosts. They can die – like bacteria they can be killed through heating and a number of other toxic agencies. They also go through 'life cycles' that involve a stage of reproduction, another basic characteristic of living organisms. The next, and perhaps most important, question is predictable: do viruses evolve through the established evolutionary mechanisms?

The answer is yes – they most certainly do.

Viral genomes mutate faster than those of any other known organism. This is part of the explanation why our immune system finds it so difficult to counteract HIV-1 once it has got inside our bodies. Within a year or two of infection there are literally billions of different evolving strains of the virus within a single infected person. While viruses do not contain their own epigenetic inheritance systems, they will sometimes take advantage of host epigenetic systems when they invade the nucleus. Are they capable of hybridisation? Again, they are the prime examples – it is the way in which new pandemic flu viruses emerge to provoke mayhem around the world. Are viruses capable of symbiotic evolution – in the jargon, genetic symbiogenesis? As I shall soon explain, they are the ultimate example of this.

Why then do some scientists insist that viruses do not belong in the tree of life? As far as I can see this appears to derive from historical reasons dating back to mistaken notions of how viruses came into being.

When life was defined in about the middle of the twentieth century, at a time when we knew a lot less about viruses, a consensus of biologists took the view that the minimum requirement was an enclosing cell membrane containing the enzymatic and biochemical means of conducting its own internal chemistry. To my mind this suggests that the definers took pains to invent a definition that specifically excluded viruses. Why should life demand a cell membrane as a defining boundary while excluding a viral envelope, which is the viral equivalent of a cell membrane? And as to the requirement for a life form to carry out its own internal chemistry, only a limited number of so-called 'autotrophic bacteria' are capable fully of carrying out their own chemistry. All other life forms, including we humans, are dependent for survival on a host of other living entities for our essential amino acids, fatty acids and vitamins. Others appear to have ruled out viruses as life forms because they are inevitably parasites – this despite the fact that so too are many different types of bacteria.

Another mistaken idea adopted at the time of the cellular definition of life was the notion that certain viruses, such as bacteriophages and retroviruses, evolved from wandering pieces of the host genome that acquire transmissible characteristics. I think that given the present understanding of viral lineages, this is no longer credible. The evidence points to bacteriophages and retroviruses evolving out of exceedingly ancient viral lineages – albeit these viral lineages, like many others, have evolved in an intimate symbiotic interaction with their hosts – what virologists term 'co-evolution' – throughout the aeons. At the time of the original definition biologists had no knowledge of the make-up of genomes. Now that we do have this information there is a very simple way in which we can put this idea to bed once and for all. If phage viruses and retroviruses were truly offshoots of the host genome, the viral genome would largely consist of similar genes to the host genome. Instead we find the very opposite – the majority of viral genes are exclusively found in viral lineages. Viruses are incredibly creative evolutionary entities, capable of manufacturing new genes all by themselves. And where there are genuine genetic commonalities between virus and host, the exchange of genes is far heavier in the direction from viruses to their hosts.

AIDS is the pandemic of our age. The causative virus, HIV-1, is a retrovirus. Even among the viruses, which have many strange and curious members, the retroviruses are remarkable. As the 'retro' of their name suggests, they contradict the now-outmoded dogma of the inexorable progression from gene to protein, via messenger RNA. Not only do retroviruses have a genome that is based on RNA rather than DNA, they also have their own enzymes capable of converting the viral RNA to its complementary sequence of DNA before they inject their converted genome into the host cell's nucleus. This is also the key to understanding how retroviruses are capable of changing the evolutionary history of the hosts that they infect. To put it in evolutionary terms, retroviruses can invade their host germ lines and thereby enter into genetic symbioses with their hosts through the creation of a new holobiontic genome – one that, in our case, is made up of a symbiotic union of retrovirus and the human genome.

HIV-1, the main cause of AIDS, spreads by unprotected sexual intercourse, whether anal, vaginal or oral, when the virus finds a way through the surface tissues. It can also enter the blood stream directly when people share contaminated injection equipment, and also from a mother to her baby during pregnancy, birth or through breastfeeding. Even at this epidemic stage, when the virus is behaving as a selfish genetic parasite, a symbiotic pattern of evolution has already begun. An important international research investigation has shown that that rate of disease progression in infected people is linked to subtypes of a human gene, known as HLA-B. This is one of the genes that determines immune responses and tissue during organ transplants. The distribution of HLA-B subtypes in the human population changes the evolution of the virus: meanwhile the virus, through lethality for specific subsets of the same gene subtypes, changes the human gene pool.

Just as we saw with hummingbirds and flowers, viruses and humans are changing one another's evolution. This is the pattern one would expect in a symbiotic evolutionary situation.

It doesn't imply that the virus is not also evolving selfishly, any more than it implies the same for the human population. At one and the same time, natural selection is operating selfishly in virus and in human but it has also begun to act at the level of the partnership. Virologists term this pattern of parasitic interaction a 'co-evolution'. From a symbiological perspective, we are witnessing how symbioses often begin with parasitism but the evolutionary situation can progress, in some cases, to one of mutualism.

The HIV-1 virus selectively hunts down an immune cell known as a CD+T helper lymphocyte. This cell has a key immunoglobulin type of chemical on its surface membrane, known as CD4, which allows the viral surface envelope to fuse with the cell membrane. The viral genome now enters the cell nucleus where the virus's own chemical enzyme, known as 'reverse transcriptase', copies the viral RNA genome to its DNA equivalent, and this, with the help of another viral enzyme, known as 'integrase', integrates the viral genome into the cell's nuclear genome. This remarkable virus–host genomic fusion is an essential step before the virus can instruct the host genome to manufacture daughter viruses that will spread to other cells, and repeat the process; meanwhile the virus spreads widely through the blood stream and tissues of the infected individual.

We note in passing the importance of the retroviral capsular envelope in evading the host immunity, then finding the target cell and fusing with the cell membrane to allow the virus to invade the host cell. As part of this process of spread, the virus will again make use of the envelope to evade the human immunity that is trying to fight it, at the same time infecting and killing more and more CD4 cells. As the disease progresses, the virus reaches the stage where billions of new viruses are created every day, meanwhile these daughter progeny are also mutating, through copying errors, at an extraordinary rate. It is this vast production and simultaneous mutational evolution of the virus within its infected host that makes it so difficult for the human immune system to defeat the virus without medical treatment. And during this proliferation phase, the virus will also preferentially find its way to the gonads, the ovaries and testicles, and it will find its way to the glands that make seminal fluid, vaginal secretions and saliva, to maximise its potential for spread to other hosts.

In the same way that a retrovirus is capable of inserting its genome into the CD4 cells, many retroviruses have the astonishing potential of inserting their genomes into the germ line of their infected hosts, the ovum and the sperm. We are observing this happening right now in a retroviral epidemic that first infected koalas in the eastern side of Australia roughly a century ago. We witness the terrifying effectiveness of sexual transmission of this so-called 'emerging infection' at first hand, with virologists confirming that all of the animals in the north are infected, and the wave of transmission passing southwards, where, other than isolated island populations, all of the koalas are likely to be infected with the virus in time. It is causing a horrific wave of mortality, from leukaemia and lymphosarcoma. But though biologists were initially worried that the retroviral epidemic might cause the extinction of the Australian koala, it is now unlikely that this will happen.

Already the retrovirus is inserting into the germ cells of the koala, so that living koalas have anything up to 40 or 50 viral loci in their chromosomes, which will now be passed down as part of the inheritance of future generations. Since this holobiontic genomic union is taking place within the nuclear genome, unlike that of the mitochondrial, the koala retrovirus inserts will be inherited in classical Mendelian manner.

To date, HIV-1 has not been seen to invade the human germ line. Some virologists had believed that this would prove impossible because HIV belongs to a subgroup of the retroviruses, called lentiviruses, which were not known to 'endogenise'. But recently lentiviruses were found in the germ lines of European rabbits and Madagascar lemurs, the latter a primate. Whether HIV will eventually become part of us remains to be determined. A multitude of other retroviruses have entered the human and pre-human primate germ lines, to contribute to the evolution of the human genome in this way, so that roughly 9 per cent of our human genome is now made up of retroviral DNA. Retroviruses that have invaded the genomes of their mammalian hosts are known as 'endogenous retroviruses' or ERVs, as opposed to free-ranging infectious viruses, which are known as 'exogenous' retroviruses. Our human endogenous retroviruses are known as human endogenous retroviruses, or 'HERVs'. HERVs comprise between 30 and 50 families, depending on definition, and these families are further subdivided into more than 200 distinct groups and subgroups. The most recent of these lineages to invade the pre-human genome, known as HERV-Ks, include ten subtypes that are exclusive to humans.

Each of these HERV families and sub-families appears to represent an independent genomic colonisation event – and therefore a genomic invasion during a historic retroviral plague that infected our ancestors. Given what we have seen of AIDS and the koala retrovirus epidemic, it suggests a grim story of ancestral survival through epidemic after epidemic. When two different sets of scientists recreated the likely original genome of our most recent human retrovirus invader, the human endogenous retrovirus HERV-K, they discovered a highly infectious exogenous retrovirus with pathogenic potential in tissue cultures. It's salutary to reflect that we are the descendants of the survivors. But now we need to consider the consequences of retroviruses entering the evolving human genome.

When a retrovirus invades a germ line cell it does so as a selfishly driven parasite. The host genome will fight back against the alien invader. This battle will continue, even if the defensive weaponry must change, when the viral genome has colonised the germ line to create 'viral loci' scattered throughout the chromosomes. Antibodies are no longer effective here within the genome but other measures, aimed at shutting down the viral loci, will come into play. One such measure is 'epigenetic silencing' (I shall explain more about this in a subsequent chapter). But such epigenetic measures as 'methylation' of the viral locus are not a permanent solution to suppressing an infectious pathogenic virus. Permanent silencing will require mutations, whether through damage to the viral genes and regulatory regions, or through the insertion of an unwanted genetic sequence into the viral genome. Meanwhile, the continuing presence of viral genome within the host germ line, often in many copies distributed throughout the chromosomes, introduces a new possibility for symbiotic genetic interaction between the two very different genomes. Over the fullness of evolutionary time, many such opportunities will arise.

We should recall that while virus and host are separate evolutionary entities, with very different evolutionary pathways, they are not unknown to one another. In fact they share an intensely interactive parasitic history. During this history the virus has evolved many different strategies for manipulating host immunity and cellular physiology, strategies in which the viral envelope, coded by the viral *env* gene, has played an important role. Meanwhile the human genome, and in particular its protective immune systems, some innate and some highly changeable and adaptive, has also evolved many strategies for hunting down and disabling the virus, its alien proteins and its alien genes.

Within the HERV loci – which comprise whole viral genomes embedded in the human chromosome – many viral genes have been silenced over millions of years by mutations. This led an earlier generation of geneticists to dismiss all viral components as 'junk DNA', but now we know that many viral loci have remained 'active', in a number of different ways. Retroviruses have their own regulatory sequences, known as 'long terminal repeats', or 'LTRs'. In the viral loci embedded in the chromosomes, these are bordering stretches of DNA enclosing the viral genes. Retroviral LTRs are regulatory dynamos capable of taking over the bureaucratic control of nearby human genes. They are also capable of interacting with other genetic sequences, including epigenetic and regulatory sequences. We also know that the huge chunks of the genome known as LINEs and SINEs are structurally related to HERVs and, as shown by Professor Villarreal, they appear to work in a complex coordination with the HERV component. Between them, the HERVs, LINEs and SINEs account for some 45 per cent of our human DNA. This begs some important questions. What role has this vast retroviral legacy played in the holobiontic evolution of the human genome? What role is it playing in our human embryology, our day-to-day physiology, including our susceptibility to diseases?

*

In 2000, a year before the draft human genome was released, Dr John M. McCoy and his colleagues in the United States and

Dr François Mallet and his colleagues in France discovered that a human protein they called 'syncytin' is coded by the envelope gene of a retrovirus locus, called ERVWE1, which is embedded in human chromosome 7. We might recall that this gene codes for the envelope protein that not only coats the virus with a kind of protective membrane but also plays a key role in the ability of the virus to find and penetrate the host target cell membrane, meanwhile evading and outwitting the complex wiles of the human cellular barriers and our white cell and antibody immune defences. Not only is syncytin coded by the viral envelope gene, or *env*, its expression is controlled by the virus's own promoter within the viral regulatory LTR. In other words, the viral locus is functioning as a viral genomic unit within the overall human genome. The viral protein, syncytin, does not code for an enzyme or structural protein, as many of our human proteins function. Syncytin changes the fate of the cells in the placental interface between the maternal and foetal circulations, so that a cell called a 'trophoblast' is turned into a 'syncytiotrophoblast'. This enables the human placenta to create a fused multi-cellular membrane, called a 'syncytium', that acts as an extremely fine filter between the maternal and foetal circulation, so that nutrients from mother to foetus and waste products from foetus to mother are obliged to pass through cellular cytoplasm. This helps to create the most deeply invading of all known mammalian placentas as well as being the finest barrier – microscopically thin as only a single cell layer can be.

Further research confirmed that the original retrovirus, which endogenised into our genome as the locus ERVWE1, invaded the primate cell line roughly 30 million years ago. Since it arrived before the divergence of the evolutionary lines of the great apes from a common primate ancestor, we humans share the ERVWE1 locus, and its placental function, with chimpanzees, gorillas and orangutans. This is why there is no 'H' for 'human' in the locus name. Were it exclusive to humans, we would add the H, so the viral name would now read HERV-WE1. And that tells us that the virus is a member of the HERV-W group. Today we also know that the virus inserted itself some 650 or so times into the genome. But the remaining 649 viral loci, spread over many different chromosomes, have all had their envelope genes switched off through mutations under the influence of natural selection. This is not merely necessary as a precaution against unwanted invasive viral emergence, it is also vitally important to avoid a conflict of envelope gene expression in such a pivotal role as reproduction. I would go further to insist that it must be this way because the ERVWE1 locus, and its expressed syncytin gene, must be recognised as 'self' by our human immunity throughout our lifetime.

Within a few years of the discovery of syncytin, two more viral loci were identified as contributing their expressed proteins to placental structure and function. The virus HERV-FRD was found to code for a protein designated 'syncytin-2', which appears to help protect the foetus from maternal immune attack through the placental barrier. ERV3 was found to complement the cell fusion role of syncytin, now redesignated 'syncytin-1'. In time a fourth virus, a member of the human-associated HERV-Ks, was found to contribute to placental function. It seems likely that there is a protective overlap in function between the four retroviruses since a small number of the population have a mutation in the ERV3 *env* gene yet they appear to be protected from sterility by the overlapping cover of the other viral genes.

Today we can list at least twelve different viral loci that contribute in one way or another to human reproduction. We still don't know what some of these viruses do, but one of these expresses its gene if a mother has a Caesarean delivery, while another expresses its gene during a natural birth. Further research on the viruses HERV-FRD and ERV3 also confirms the same pattern of mutational suppression of all potentially rival loci of these viruses scattered throughout the human chromosomes. Only the selected loci can be regarded as self.

In May 2012 it was my pleasure to make a trip to the historic Uppsala University, in Sweden, famous for its association with Carl Linnaeus, who devised the system of the classification of life still used by biologists. I came here at the invitation of my friend and colleague, Erik Larsson, Professor of Pathology at the University Hospital, who is an international expert on human endogenous retroviruses. I was already aware that Professor Larsson and his colleagues had conducted important pioneering research on the role of HERVs in human embryology, and on their contribution to important aspects of normal physiology, particularly in relation to placentation and human reproduction. I wrote about this extensively in my book, *Virolution*.

Professor Larsson's research pointed to a major role of HERVs in human evolution as well as in human diseases, such as cancers and the autoimmune diseases. To get a better understanding of both these roles, we needed to know what viral genes might be contributing to normal human embryology and physiology. In particular we needed better techniques of searching for viral gene expression in different human cells, tissues and organs. Until recently scientists had been limited to looking for the expression of viral genes as messenger RNA. This had given us important clues as to what was happening at tissue level, but we needed to develop accurate deep sequencing techniques that would allow us to show viral proteins at work in the human cells. This has been the focus of research in Uppsala for many years.

During my visit, it was also my pleasure to meet Professor Larsson's colleagues, the molecular biologists, Professor Fredrik Pontén and Dr Per-Henrik Edqvist of the Rudbeck Laboratory. One of the great mysteries of human embryological development is how, from the original pluripotent cells of the fertilised ovum, all the different cells that make up the various tissues and organs arise. For many years, Pontén and Edqvist have been working with other Swedish molecular biologists in exploring this mystery. In particular they have screened the cells of different human tissues and organs to see how the expression of proteins might differ between, say, a nerve cell and a blood cell, or a cell from the liver or kidney. This enormous undertaking, which is a natural follow-up to the Human Genome Project, is known as the 'Human Proteome Project'. At the time of my visit, it was nearing completion and it had provided fascinating insights into how the day-to-day machinery of the different tissues and organs work. In essence they arrived at two related conclusions: each specific tissue cell had a small number of proteins that were exclusive to the cell type, on average perhaps six or seven, but the major difference between the cells of the various tissues and organs was in the variations of the overall expression profile of a very wide range of proteins that were common to most.

Pontén and Edqvist also cooperated with Professor Larsson and his Department of Pathology, in looking for the expression of HERV-derived proteins. To do so they devised a new system involving extensive tissue screening with a pair of antibodies raised against protein sequences derived from two different sections of HERV envelope gene. The Uppsala-based scientists now extrapolated this system to study three retroviral loci in the human genome, ERVWE1, ERV3 and HERV-FRD, looking for significant expression of their envelope genes in a wide range of human cells, tissues and organs. What they discovered was original and astonishing. These virus-derived envelope proteins, pre-evolved to interact at a deep level with our human physiology, were being expressed at a significant level – one that suggested physiological function – in many different human cells, tissues and organs other than the placenta, including the brain, the liver, the bowel, skeletal muscle, the heart, the skin, the adrenal glands, the salivary glands, the insulin-secreting islets of Langerhans in the pancreas, and the testis, as well as in multinucleated inflammatory cells found in the blood as part of the reaction to foreign invaders, such as bacteria and infectious viruses.

This was powerful supportive evidence for virus–human symbiosis at genomic level. In each case the viral envelope gene was under the control of the viral promoter sequences, which were conserved by natural selection within the virus's own regulatory LTRs. It seems likely that these pioneering Swedish colleagues, and others, will extend this type of study to other viral loci within the human genome. We still have a great deal to learn about what such viral proteins might be doing in these many different tissues and organs, both in terms of normal physiological function and in terms of disease.

So-called because it shows up as a starry shape within the substance of the brain, the astrocyte is a support cell involved in local immune responses within the brain and central nervous system. The Swedish scientists confirmed that syncytin-1, the envelope protein of ERVWE1, is normally expressed in modest amounts in these cells. Other scientists in France, Italy, Germany and America have discovered that the same viral protein appears to be over expressed in the local astrocytes within the disease-affected parts of the brain and central nervous system in patients with multiple sclerosis. Italian scientists have also shown that during the first acute attack of MS, a virus closely resembling ERVWE1 appears in high levels of intensity in the blood. In other laboratory tests, it has been shown that the affected astrocytes secrete a chemical that is lethal to a type of brain cell called the oligodendrocyte, which manufactures myelin, the substance that coats nerve cells like the insulation on an electrical cable. Damage to myelin is the central pathology of MS. This same virus falls to unmeasurable levels in patients who respond well to beta-interferon therapy. Could it be that a defective regulatory control of the ERVWE1 viral envelope gene expression in the astrocytes is playing some role in the pathology of MS?

While the evidence for a causative role is not yet strong enough to be definitive, the possible association between this virus – now labelled MSRV/HERV-W – and MS is undergoing extensive further testing, so that in time we shall have an answer to this important question – and perhaps to more of the questions that are beginning to accumulate about the potential viral contribution to various cancers as well as a wide range of illnesses that are included under the umbrella description of the autoimmune diseases.

*

In 2001, when the first draft of the complete human genome showed that roughly 45 per cent of the human genome appeared to be made up of retroviruses, or virus-like entities, such as LINEs and SINEs, some biologists dismissed this huge genetic inheritance as junk, the graveyard of past viral infections. But today we have become a good deal more cautious in our interpretations. The 2001 papers in *Science* and *Nature* had shown that some 50 per cent of the genome was accounted for between the protein-coding genes and the various virus-related sections, but the papers had also revealed that approximately 50 per cent of our DNA appeared to code for nothing that we recognised at this time. Of course, some biologists once more labelled it junk, but others were now more cautious. This new mystery would become the *raison d'être* of a deliberate investigative enterprise, inspired by the very shock of our exposed ignorance beyond the tiny 1.5 per cent fraction of protein-coding genes. This investigation, involving a consortium of research groups worldwide, would be encouraged and funded by a public research project launched by the National Human Genome Research Institute in the United States in September 2003. Its acronym was derived from the very nature of the mystery: the 'Encyclopaedia of DNA coding elements' – ENCODE.

twelve

Genomic Level Evolution

The fact that selection can work simultaneously on both genetic and epigenetic variation complicates matters even further . . . models that incorporate the effects of epigenetic variations . . . show how [this] leads to different evolutionary dynamics.

EVA JABLONKA AND MARION J. LAMB

If we look at what happens when a fertilised egg develops into the complex wonder of the human baby, logic would tell us that the process of embryology must be directed by an integrated and coordinated system of control. The fertilised ovum, or 'zygote', is a pluripotent cell – a cell that can develop into any of the organs and tissues that make up the individual human being. When the zygote first begins to divide, the daughter cells retain this pluripotency throughout the very early divisions. If cells separate from the embryo at this stage, each cell is still capable of giving rise to a complete healthy individual. This is how identical twins, triplets or quadruplets usually arise. But soon the developing mass of cells develops into two different entities, an encircling hollow ball of cells that will become the placenta and an inner cell mass that will become the foetus. At this stage the symbiotic endogenous retro-viruses kick in and express their envelope proteins, which help to create the deeply invasive placenta that, in a quasi-parasitic pattern, burrows into the maternal uterine wall and constructs the fused cell interface between the maternal and the foetal circulations. While this is happening, a complex array of signals takes over the control of the inner cell mass, instructing the cells to divide and multiply, but also recognising very early the need for the cell types to change, so that selected embryonic cells begin to change into the forerunner cells of the different tissues and organs.

Here we face the enigma that all the cells in an organism have the same DNA, which incorporates the same sum of genes. Therefore different organ and tissue cell types must be determined by mechanisms other than the sum of all the genes they contain. From the studies of the Swedish scientists described in the previous chapter, we now know that the difference between a brain cell and, say, a kidney cell or a circulating blood cell, comes mainly from the profile of expression of genes within the cell. We also recall that each tissue-type cell also expresses a limited number of genes that appear to be particular to that cell – perhaps about half a dozen for each cell type. This fate of cells, and the organisation of these cells into the growing complexity of form and function that will make up the different tissues as organs of the head and body parts, is controlled by what geneticists call the epigenetic system.

Some readers might become a trifle worried at this point, since there appears to be a general notion that epigenetics is immensely complicated. It is even fair to say that the world of epigenetics seemed confusing to scientists until recently – but the reason for this was that the definition and scope of epigenetics were under-going a rapid evolution in themselves. As our understanding has grown, the basic principles have, thankfully, become much easier to grasp. In particular, our growing understanding of the so-called non-coding RNAs has clarified things so that not only can we redefine epigenetics in simpler terms but we can also see how it provides a fascinating answer to the remaining mystery of that huge unknown section of our human DNA.

The epigenetic system is essentially a system of regulatory control of the functioning of the genome. It comprises a number of different mechanisms that act in an integrated and coordinated manner to control the activation and closing down of genes. But its role also extends beyond genes to work in what might be compared to a housekeeping and regulatory coordination affecting the entire genome. The easiest way to understand this is to examine how the various mechanisms operate.

We have seen how a gene is the genetic sequence that codes for a protein, or perhaps more correctly, a parcel of different proteins. Our genome contains roughly 20,500 such genes. We have also seen how a specific cell type, and thus the make-up of the different tissues and organs of the body, is determined mostly by the profile of expression of a large number of genes. The epigenetic system decides that profile of genes, controlling which genes switch on, when they do so, what quantity of protein they express, and so on. Before we examine how it does so, I would like to explain something interesting about this epigenetic system of control.

The DNA that codes for genes and the other functional genetic sequences of the genome is fixed when the parental germ cells unite to form the fertilised cell, or zygote. This aspect of your genome remains exactly the same throughout your life unless changed by a mutation or an invading virus. But your epigenetic system, and its regulatory effects, is not fixed throughout your life. It is capable of changing its regulatory commands, for example by responding to signals coming from your internal physiology, and even through signals coming from your environment. In plants, for example, it is the epigenetic system that tells them that spring has arrived. And in animals, including humans, the epigenetic system is similarly sensitive to important changes in your living circumstances, such as the impact of disease, protracted stress, severe pain or starvation. In other words, although your genes stay the same throughout your life, the expression of those genes in your various tissues and organs can, and will, change because of signals arriving into the systems of epigenetic control. The implications go even deeper: your epigenome is capable of learning from experience and changing to accommodate that experience. More extraordinary still, those changes will sometimes be inherited by future generations of offspring through mechanisms known as epigenetic inheritance systems.

The inheritance of epigenetic changes through epigenetic inheritance systems means that epigenetics has evolutionary potential: this is why I included it as one of the mechanisms of 'genomic creativity'. Moreover, the potential for outside influences to bring about epigenetic change offers exciting potential for medicine. For example, it might lead to future medical therapies aimed at changing the expression of disease-causing genes.

Today we recognise four main epigenetic control systems, which can be seen in the figure below:

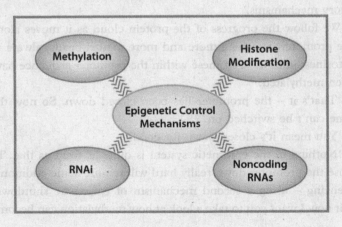

Would you like to join me in a new journey on our metaphorical train? We find ourselves reversing along the twin track of the DNA of a gene, making our way back past the first exon, to arrive at the nearby stretch of DNA that is the 'promoter' – the region that switches the gene on and off. As before, we hop down so we can observe what is happening as one of the epigenetic mechanisms swings into operation.

We hear a deep buzzing sound nearby and are then startled as a small cloud looms into view, buzzing like a bee. The cloud is a protein called a DNA-methyltransferase. We see that it is bearing tiny clusters of atoms, resembling chemical beads – these are 'methyl' chemical groups, made up of a carbon atom attached by covalent bonds to three hydrogen atoms. As we watch, the cloud attaches a methyl bead to a nucleotide in one of the sleepers.

'Go ahead – check which nucleotide.'

'It's a cytosine – a C.'

'Okay – so with the methyl chemical tagged on to this it is now a methylated cytosine. Believe it or not, this simple chemical change is all there is to one of the most powerful epigenetic regulatory mechanisms.'

We follow the progress of the protein cloud as it moves along the promoter, attaching more and more methyl beads, always to cytosines, until most of these within the promoter sequence have been methylated.

'That's it – the promoter has been closed down. So now the gene can't be switched on.'

'You mean it's closed down for good?'

'Nothing in the epigenetic system is quite as fixed as that. To close the promoter down really hard will require some additional silencing – using a second mechanism of epigenetic shutdown. But now I want you to take a look at how methylation can become an epigenetic inheritance system. To understand this you need to take a closer look at the nucleotides adjacent to the cytosines.'

'You mean the other halves of the sleepers?'

'No – we already know that they will be guanines. Because cytosine always binds to guanine. Look at the adjacent sleepers.'

It takes you a minute or two, because you don't see the pattern until you have examined half a dozen or so.

'There always seem to be guanines adjacent to the methylated cytosines.'

'That's right. These cytosine-guanine "couplets" are the key to how the methylation status of genes is inherited to new generations. If we now climb back aboard our train we can actually watch it happen.'

In the blink of an eye we find ourselves entering the genome of a germ-forming cell that is in the process of copying its genome into a sperm or an ovum. We hop down again so we can watch what is happening to another promoter region as it is in the process of being copied. And here we notice something very interesting take place.

To begin with, I draw your attention to one of the C-G couplets. We can hardly miss the fact that the C is methylated – it has its smoky bead attached. We watch as the double helix cleaves apart, and then the process of copying begins, with nucleotides being matched with one another and the new rail forming.

We follow the copying from the 'sense' strand to the new 'anti-sense' strand and, with a chuckle of recognition, we observe that wherever there is a C-G couplet on the sense strand, it creates a mirror image copy of G-C in the daughter strand. I bid you wait so we can witness another surprise.

With amazing speed, we observe the arrival of another buzzing cloud, which appears to notice the unmethylated couplets on the daughter strand that stand out in contrast with the methylated couplets opposite on the maternal strand. With that same efficiency as before, it moves along the daughter strand, methylating every complementary couplet on the daughter line.

'The methylation status has been transferred?'

'We have witnessed the operation of an "epigenetic inheritance system" which is a way in which a change in the methylation code can be inherited by future generations. What's more, our train analogy has allowed us to get close enough to witness something definitive. We have seen how methylation can switch off the expression of a gene. If inherited by future generations, this would have evolutionary potential since it would change the heredity of those future individuals through altering gene-profile expression. In other words, methylation status can bring about hereditary change. It's a force within my umbrella of genomic creativity. Yet, as we see, it comprises adding a simple chemical to a pre-existing nucleotide, cytosine. There is no actual change in any DNA sequence. Every other force of genomic creativity that we have seen to date – mutation, genetic symbiosis, hybrid-isation – acts through changing DNA sequences, yet this epige-netic mechanism changes heredity without changing genes. An epigenetic inheritance system is a genomic force but it is not a genetic mechanism. This is why I coined the term "genomic" creativity rather than "genetic" creativity.'

Methylation is a very important epigenetic mechanism during the formation of the embryo in the mother's womb. The vitamin folic acid plays an important role in this process of methylation during early embryological development. This is why a lack of folic acid in the mother's diet in those early months of pregnancy can damage the foetus and increase the propensity for developing spina bifida. Extensive defects in the methylation patterns throughout the genome is also a feature of many forms of cancer, a finding that is being extensively investigated in the hope that it will provide enlightenment and potential avenues of treatment in the future.

Another situation where methylation status may be important is morbid obesity, with its tendency to maturity onset diabetes. Some studies have shown that epigenetic factors, particularly changes in methylation status in key areas of the genome, may be playing some part in this.

'Can we do nothing to help ourselves?'

'Yes, we can. Unlike the fixity of genes, epigenetic regulatory systems are amenable to change. And something as simple as regular exercise can change things back to a healthier epigenetic code.'

'But, hold on! What's happening? We appear to be moving again.'

'Time for another trip. I want you to observe another epigenetic mechanism as it actually happens. But this time we are going to restore the double helix in all of its splendour.'

We watch as our train moves away from the genome in the ultramicroscopic landscape, sufficient to observe the spectacular beauty of DNA's natural twist.

'This time we need to observe the actual structure of the chromosome – in this case human chromosome 6, the chromosome that contains that supremely important Major Histocompatibility Complex. And our first surprise will be to discover, in passing, that those early geneticists who made life difficult for Oswald Avery may have had a point when they insisted that proteins had something to do with the mystery of the gene.'

*

We discover to our delight that our magical steam engine is capable of hovering within the ultramicroscopic landscape at sufficient distance for us to observe the double helix grow small enough to appear as a fine thread in the distance – far enough away for us to notice things that were not apparent before.

Not only does the incredibly long molecule of DNA spiral within its molecular structure, the twin track then coils for a second time in a broad spiral around some strange globular structures, that from this distance resemble tennis balls. The tennis balls are proteins, called histones, and they are packed together as structural units of eight balls, in racks of four on four. These eight-packs are themselves wound around a central spine of another linear protein, a central spine unlike the phosphate spines of the DNA molecule. This cruder secondary spiral, of the DNA thread winding around the eight-packs of histones, extends the entire length of the chromosome, winding away into the fathomless distance.

You appear to be nonplussed.

'It is extraordinary to glimpse the gargantuan wonder of this secondary chromosomal structure – if my mental arithmetic is anywhere near accurate, chromosome 6 is something like 150 million nucleotides long.'

'Where are we headed?'

'To the stretch that codes for the Major Histocompatibility Complex.'

Now you turn to me with a new question on your lips: 'What's this new tip about?'

'We're going to take a look at a second epigenetic system, called the histone code. And like the methylation status, it's really very simple.'

You look a trifle dubious.

'It's all about things the epigeneticists call histone tails.'

The engine chugs closer to a single section of the chromosome, at a place where we can examine the structure where the eight-packs of tennis balls are inclined to us, side on. These appear to be packaged together like newly harvested onions, bound tight to the central string of the spine.

The tight packaging of the histone eight-packs suddenly opens up. We watch how the thread of DNA loosens up as the individual eight-packs are now teased out into a looser arrangement.

I edge the engine closer. 'Look more closely at the eight-packs.'

'There's something poking out . . . They're sporting tails.'

'Chemical tails – yes.'

Since the histones of the eight-packs are proteins, they are made up of long strings of amino acids. These tails, trailing from the histones, are side chains of amino acids. The key to understanding is that these amino acid tails poke out beyond the broad spiral of DNA thread that is wrapped around the histones. As we watch, one of those buzzing protein clouds hoves into view, hauling one of those chemical beads we saw before. We watch in silence as it attaches the bead to one of the dangling tails. Immediately the whole arrangement begins to change again. The loosened structure of the histone packs begins to tighten into the onion pack again.

'It's coming together.'

'The histone proteins are exceedingly sensitive to the attachment of certain chemical molecules to their tails.'

'Like moths to pheromones?'

Now you have me chuckling with you. 'Sometimes it's our old friend, the methyl group. But it can be an acetate group, or a phosphate – there's a range of different simple chemical groups that can trigger the change. And the change can only be one or the other, a local loosening up of the histone packs, or a tightening up. The chemicals attach to specific amino acids within the histone. For example, it might be the adding of an acetyl group to the amino acid lysine, or the adding of a phosphate group to serine, or the adding of a methyl group this time not to cytosine of the DNA but lysine on the histone tail.'

Together we watch how the tightly wound spiral of DNA thread that is wrapped around the tightly packed histones loosens up again.

'So you know what has just happened?'

'The gene – or whatever sequence is coded by this stretch of the DNA – is closed down when the thread is packed up tight.'

'And ready for translation when it uncoils – exactly!'

'So the histone code switches a gene on or off, just like the methylation status?'

'It may look very simple, but there is nothing remotely accidental in the acetyl, phosphate or methyl groups cosying up to the tails. It is under a very careful control by other elements within the epigenetic control system that would make the secret police force of a dictatorship look like amateurs. And just like the methylation status, the histone code is also amenable to change within the lifetime of the individual. It is responsive to stimuli entering the genome, through environmental influences. And, like methylation, it has the potential to change heredity, and thus bring about evolutionary change, without changing the DNA of the genetic code.'

'Just how powerful,' you ask, 'is this histone code?'

'Let me give you a single example. That protein cloud we saw is actually an enzyme called a "deacetylase". What it did was remove the acetyl chemical group from the tails of a single histone pack. Its name, in the jargon, is deacetylase HDAC11, and we watched it switch off a gene that codes for a protein involved in the body's immune system. That protein decides whether you or I will respond to a certain antigen as self or foreign. In medical terms, this single "epigenetic mark" will influence in an important way our future immune tolerance – in other words, how we might respond to a dangerous invading microbe or how, if we suffered an organ failure, we might react to an organ transplant.'

I can explain through another example. Identical twins – in the jargon 'monozygotic twins' – are conceived as clones of one another. Thus they are conceived, and live out their lives, with identical genomes. The sum of all of the epigenetic systems in the body is known as the 'epigenome'. Identical twins are *conceived* from the same pluripotent cells, so they begin as embryos with identical epigenomes. We formerly believed that this implied that identical twins were also *born* with identical epigenomes, but now we know that this is not true. The epigenome of every foetus, including those of identical twins, has already begun to change by the time of birth in response to environmental influences arriving into the physiology of the foetus during development in the womb. Of course it doesn't stop there. A study in Spain showed that, depending on the circumstances in which each of two identical twins grows up and lives out their lives, they continue to accumulate these epigenetic differences.

In practice, the epigenetic silencing of genes through methylation will often be reinforced by a silencing histone code being applied to the same gene's promoter – a belt and braces guarantee that the gene remains silent.

Perhaps we should take a breather. I want you suitably rested before we confront the newly discovered and even more extraordinary epigenetic story of what some geneticists formally, and somewhat disparagingly, called DNA's Cinderella sister – that second nucleotide molecule called ribonucleic acid. In short, RNA.

Of course, we have already come across RNA. Outshone by its stellar sister molecule, DNA, we rather assumed that RNA had had its day in some ghastly early model of the Earth, a time before the greening of the planet, when life was still bogged down in a chemical stage of its evolution, with self-replicators competing with one another for the grubby chemicals they needed in the dirt of the primeval planet. It was understandable, in retrospect, that amid the fabulous catalogue of discoveries deriving from the discovery of DNA and how it coded for proteins, scientists thought of genes, and the human genome, more or less exclusively in terms of DNA as the master molecule.

Today, however, we recognise that this was a blinkered vision – and it was this same blinkered vision that led to half the human genome remaining blank in that pie chart dating back to 2001. The solution to that enigma lay in the more recent discoveries of extraordinary new roles for RNA in the burgeoning new discipline of epigenetics. This is changing much of what we formerly assumed about genetics, biology, molecular biology and medicine. These new discoveries are so bracing and challenging that we are obliged to rethink our views on how the genome works. This dilemma is throwing up some very fundamental questions. What, for example, do we really mean by a gene? If we adhere to the concept of the gene as the unit of heredity then we shall have some redefining to do. For example, Thomas Gingeras, one of the investigating experts involved in the ENCODE Project, goes so far as to argue that the fundamental unit of the genome – the basic unit of heredity – should no longer be the gene at all but the RNA transcript decoded from DNA.

thirteen

The Master Controllers

> *We had many discussions on DNA, for I had come to Oxford*
> *with two half ideas, both of which were more than half wrong.*
>
> SYDNEY BRENNER

This new chapter of discovery began as long ago as 1991 with a discovery by American biologists Victor Ambros, Rosalind Lee and Rhonda Feinbaum when they were studying a single gene, called *lin-14*, which regulates development in the worm, *C. elegans*. We might recall that this tiny worm was the experimental subject chosen by Crick's friend and colleague, Sydney Brenner, for his pioneering experiments into the genes and molecular biology of development. We might also recall that this tiny worm proved so helpful and amenable to Brenner and his colleagues that it subsequently became the test organism for thousands of laboratory experiments all over the world. Brenner's own studies were extended by the biologists Robert Horvitz, in the United States, and John Sulston, in England, where their efforts were crowned by the Nobel Prize in Physiology or Medicine in 2002. In the press release from the Nobel Institute, the award was made for their discoveries in the 'genetic regulation of organ development *and programmed cell death*'.

The italics are mine, because I want to draw attention to what those words might imply.

In an adult human being more than a thousand billion cells are created every day through cell division, or 'mitosis'. In every such cell division the entire genome is copied. At the same time an equal number of cells die through a form of controlled suicide. This is what is referred to as 'programmed cell death', or 'apoptosis'. It is amazing, when we think about it, that death as well as life is programmed into our genome; and Brenner's work led to our first understanding of the genetics involved in bringing about death. Specific regulatory genes, and genetic pathways, are involved in this darker side of programming. And yes, RNA – that strange, almost quixotic, sister molecule – the Cinderella of the nucleotide sisters – is involved in this curious regulation.

A variety of tiny RNA molecules, between 20 and 30 nucleotides long, was discovered back in 1991, but scientists were unsure what they represented. A few years later, the UK-based botanist David Baulcombe, together with his colleague, Andrew Hamilton, found that some small interactive RNA molecules, or siRNAs, were somehow capable of silencing messenger RNA molecules. In the United States two geneticists, Craig C. Mello and Andrew Z. Fire, adopted the *C. elegans* model to study this in finer detail. Focusing on the genetic control of a muscle protein that was important in the worm's normal sinewy movement, they injected the gonads with siRNA molecules and watched how it affected the worm's movement. To start with, they broke down the double-stranded siRNA into its two strands, known as 'sense' and 'anti-sense', first testing the sense strand – the RNA that exactly matched the original genetic DNA coding – and then the anti-sense strand. Neither had any effect on the worm's movements. Only when they injected both the sense and anti-sense RNA at the same time did something happen. The worm developed an abnormal twitch – the same dysfunctional movement that they saw when the relevant gene was damaged by a mutation.

This led to the startling realisation that these small RNA molecules were capable of taking out specific messenger RNAs. In other words, even after the translation had already taken place – with the messenger RNA copied from the gene, the introns removed and the exons spliced together for the final messenger RNA molecule to move out into the cytoplasm ready to code for the protein – these tiny RNA molecules would terminate the process.

In passing we realise why the Swedish scientists thought it so important to look for viral proteins not merely as messenger RNA transcripts but as the actual expressed proteins within the cells.

This epigenetic mechanism came to be called 'RNA interference', or 'RNAi'. It was yet another mechanism of epigenetic control. The implications were startling. These RNAis recognised key sequences in the specific messenger RNA molecule, so they could home in to inactivate or even to destroy it entirely. In 2006, Fire and Mello were awarded the Nobel Prize in Physiology or Medicine for their discovery.

From the earliest days of genetics, scientists had focused on what was perceived to be dogma – that genes invariably coded for proteins. Then scientists discovered that this required RNA as the messenger molecule, mRNA. It also required a different type of RNA for the transport of amino acids to the ribosomes, the so-called transfer RNA, or tRNA, as well as a third variety of RNA that was part of the basic ribosome structure, the so-called ribosomal RNA which somehow read off mRNA while translating its coding to the protein. However, in those early years these roles were perceived as secondary, or at best intermediary, to the noble Gene-to-Protein central axis – the central paradigm. But now we had a fourth type of RNA, these terminator RNAi molecules! Of course, the three different varieties of RNA, other than messenger RNA, must be coded within the nuclear-based chromosomes by sequences of matching DNA. But these coding zones, the DNA sequences that coded for these non-messenger forms of RNA, could hardly be called genes. They did not code for proteins, on the contrary, the RNA molecules they coded for were end points in themselves.

At this point geneticists were faced with a dilemma. How were they to classify the genetic sequences that coded for these RNA upstarts? And now that RNAi had been discovered, a number of other small 'non-coding' RNAs were further challenging the paradigm.

Some geneticists toyed with the idea of an 'RNA gene', a gene that coded for an end-point RNA. But others had reservations about the very idea of RNA genes. Whatever the terminological inconsistencies, there could be no doubting the fact that the human genome coded for a surprising variety of RNA molecules that did not code for proteins, but nevertheless had important roles to play in the control and expression of genes.

RNA inhibition by small, non-coding, double-stranded RNA molecules has proved not only important in theory but also of practical usefulness to biologists and geneticists, allowing them to examine the role of specific genes by observing what happens to cells or life forms when the gene is 'knocked out'. The potential for medical therapy is equally important. Some women bear the frightening burden of inheriting one or other of the breast- and ovarian-cancer-associated gene mutations, BRCA1 and BRCA2, while other patients present with the early symptoms of Huntington's disease. All it would take to alleviate the suffering and distress of such patients would be to switch off the relevant mutated genes. Some time in the future, perhaps sooner than we might imagine, molecular geneticists will find a way to do this.

Moreover, RNAis are not the sole contribution of RNA to the regulation of genes. A different group of small non-coding RNAs, known as Piwi-interacting RNAs, or piRNAs, appear to be playing an important role in the epigenetic silencing of dangerous viral sequences in the human genome. Moreover, there is another, perhaps even more astonishing class of non-coding RNAs that regulates the human genome, a relatively new discovery that explains that mysterious black hole in the 2001 draft genome – the 50 per cent of our human DNA that was left a baffling blank.

*

We mammals have evolved sexually differentiating chromosomes, the X and the Y, so that females inherit two copies of the X, one from each parent, and males inherit an X from the mother and a Y from the father. In addition we inherit 22 non-sex-differentiating chromosomes, called 'autosomes' from each parent, making up a total nuclear genome of 46 chromosomes. While the Y chromosome contains an estimated 78 protein-coding genes, largely concerned with testicular development as well as the male physique, fertility and sperm production, the X chromosome contains roughly 2,000 genes, few of which have anything to do with sexuality. This chromosomal discordance between the sexes led to a potential imbalance in regulation during embryological development. If the sex-linked chromosomes were to be fully expressed during embryo-logical development, female embryos – and females throughout life – would be subject to double the dose of the X-linked genes, while male embryos – and males throughout life – would be subject to a single dose of those same X-linked genes. This could lead to unwelcome regulatory clashes.

In 1961 Mary F. Lyon, a former pupil of the epigenetic pioneer Conrad H. Waddington, realised that a solution to this develop-mental riddle might be to switch off one of the two X chromosomes in females. Lyon was duly vindicated when geneticists subsequently confirmed 'X-inactivation' in female embryos on about the sixteenth day of embryological development. Curiously, the inactivation does not select for the X chromosome from any particular parent; it appears to choose randomly between the maternally or paternally inherited X chromosomes and it does not switch off all of the inactivated chromosome, but something more like 60 per cent of its genes. The remaining 40 per cent are important in protecting females from diseases caused by recessive mutations on the X chromosome. This is why females are rarely affected by colour blindness or haemophilia – they would need a double dose of the mutated recessive genes – but males need only a single copy on their solitary X chromosome.

In 1991, some thirty years after Lyon came up with the idea, scientists working at Stanford University discovered that a single gene on the inactivated X chromosome played a key part in the process of X-inactivation. They called the gene *Xist* after the role it encoded, as the 'X inactive specific transcript'. They also assumed it must work through translating to a corresponding *Xist* protein. But when they looked for the protein they couldn't find it. This was baffling, since they could trace the gene's expression to the relevant messenger RNA, which was spliced to remove the introns and the exons, which were cobbled together as usual. But the mRNA failed to move out to the ribosomes, where one would expect the protein to be manufactured. I think it might be timely for us to make another exploration on the train, to observe one of the most mind-blowing of recent discoveries about our human genome. As we enter the magical landscape I direct you to one of two similar tracks, running parallel to one another – the two X chromosomes. We have entered the genome of a female foetus on the critical sixteenth day of embryological development.

We watch cell division taking place within the early embryo, with replication of the genome, the double tracks of the two X chromosomes unzipping along the weak hydrogen bonds of the interlocking jigsaws of the sleepers, to liberate the sense from the anti-sense strands of DNA. The speed of copying is impressive. A blizzard is approaching, but the composite flakes are not snow but the RNA-bound nucleotides, G, A, C and U. As we continue to watch, sections of the sense strand begin to glow in different colours. It is all part of the magic that enables us to make out sequences that mark out genes, or promoters, or viral sections, or sections we currently know diddley squat about. The process is very similar to what we saw with the coding for a protein, with the stretch of DNA being copied to its matching stretch of messenger RNA, but here the copying appears to go on and on, extending far more extensively than the thousand or so nucleotides we would expect for a single gene. A huge molecule of RNA is being fashioned, comprising some 17,000 nucleotides. It appears to be peculiar also in its intrinsic structure, with the equivalent of genetic full stops, or 'stop codons', at intervals throughout its length. We have never seen anything remotely like this structure before.

'What is it?'

'It's a long non-coding RNA – the product of what some geneticists call an RNA gene. The scientific name for it is *Xist* RNA.'

We watch as the RNA molecule flows like a leaking pipe over the targeted X chromosome, changing the gene-activating histone epigenetic marks in a way that hauls together the histone packs into a tight non-coding formation and calling in methylation protein clouds to switch off the cytosine-guanine couplets.

'What's it doing?'

'It's switching everything off but not the whole chromosome, just the unwanted 60 per cent.'

Xist was duly recognised as the first of a remarkable new class of epigenetic controllers – what we now call the 'long non-coding RNAs', or lncRNAs. But soon afterwards a second, very powerful, lncRNA was discovered and it explained another epigenetic mystery.

Geneticists had already observed that the genome can recognise the specific parental origins of the matching pairs of chromosomes. For example, it could select the specific paternal, or the maternal, chromosome when allowing certain genes, or even whole clusters of genes, to be expressed. This epigenetic mechanism, which is known as 'imprinting', is a key factor in the genetic causation of diseases such as Prader–Willi and Angelman syndromes, because it selects a damaged chromosome according to a specific parent of origin even though the chromosome inherited from the other parent of origin is perfectly normal. Geneticists discovered that a key mechanism of imprinting was caused by epigenetic silencing of a whole region of the non-chosen chromosome by another long non-coding RNA, known as *Air*.

Inspired by these discoveries, scientists began to search for more of these long non-coding RNA molecules to discover that they are pervasively transcribed throughout mammalian genomes. In time, lncRNAs were duly recognised as part of a newly recognised and very powerful epigenetic regulatory system, giving rise to an explosion of new research. This exciting new venture is still taking place as I write, but already we know that our human genome, like that of all plants and animals, contains vast numbers of long and small non-coding RNAs within which the lncRNAs comprise a class of their own, ranging in size from 200 to more than 100,000 nucleotides long. And what it is revealing, in terms of the coding for these lncRNAs, is at first glance bizarre, but yet also wonderfully logical.

There is a second, utterly different, reading from the entirety of the genome. This reading is not concerned with normal boundaries of genes, or regulatory sequences. It can code from any stretch, whether remaining confined to, say, an exon, a group of exons, a promoter region, or a combination of promoter and exons, or the regulatory LTR of a virus, or all of these in a single sequence. And the transcripts are all non-coding RNA molecules.

This is the explanation for the unknown 50 per cent of the genome.

I gaze at your puzzled face, aware that we are still aboard our magical mystery train, on our way back into the normal world of these pages.

'The puzzle was an artefact caused by how they actually derived their genomic sequences. The 2001 reading of the draft full genome was based on compiling all the messenger RNA sequences found in the human cell, through a technique called "expressed sequence tags", or ESTs. The messenger RNA is reversed to its complementary DNA, known as cDNA. So the 2001 pie chart was based not on the DNA of the human genome but the sum coding of all the messenger RNA that is expressed from that DNA.'

You are still shaking your head.

'The whole genome, or most of, is actually translated twice – in two utterly different ways when it comes to genetic sequences . . .'

'Ah, the whole genome – it's copied twice.'

'Exactly. That was why the black hole was roughly 50 per cent of the genome. It was the missing second translation into non-coding RNAs.'

*

We see now how that older attitude to the genome, with its excessive emphasis on protein-coding DNA genes, blinkered our vision to the bigger picture. This more comprehensive understanding is still being further evaluated.

The nomenclature of non-coding RNAs is simple and predictable; they are named after the sequence within the genome that encodes them. So a sequence based on a single exon – or intron – is called an 'exonic' or 'intronic' lncRNA. A sequence based on a gene is a 'genic' lncRNA, and so on. The lncRNA can be derived from the sense strand of DNA, from a regulatory region, a promoter sequence, from an entire gene, including all the exons and introns, or even from the intervening sequences between different genes that includes upstream regulatory regions. It can also be coded in much the same way along the anti-sense DNA strand. Some are coded in both directions, being known as 'bidirectional transcripts'. There are even mitochondrial lncRNAs, and virus or LINE- or SINE-associated RNAs, known collectively as repeat-associated lncRNAs. Their purpose is epigenetic control of the genome, so that, as this extraordinary repertoire would suggest, lncRNAs control a great variety of different genomic functions.

One such function is to focus on what are known as regulatory proteins – proteins that switch genes on or off. The lncRNA will fix to the DNA thread at an appropriate point, grab hold of the regulatory protein and direct it to where it needs to be to influence the appropriate gene. Although the research is still ongoing, we already know that lncRNAs are involved in the epigenetic, genetic and whole genomic regulation of many different and sometimes very complicated biological processes. They appear to be important during the embryonic stage of development, where they play a pivotal role in embryonic stem cells – the pluripotent cells that make up the very early embryo. Here the lncRNAs are involved in the differentiation of these stem cells into those that will differentiate into the different tissues and organs. As we have seen with the Prader–Willi and Angelman syndromes, they appear to play important roles in the hereditary aspects of some of the inherited disorders of metabolism. They are also thought to play some role in many different cancers affecting breast, bladder, colon, prostate, lung, bone, brain, as well as melanomas and leukaemias. In addition it is thought they may contribute to the autoimmune disorders, coronary artery disease, neurological disorders such a spinocerebellar ataxia, fragile X syndrome, Alzheimer's disease, and, possibly, the ageing process.

We can now fill in the blank space to create a new pie chart of the genome:

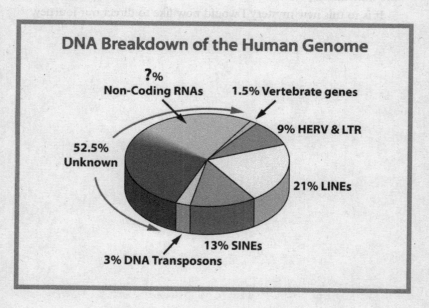

DNA Breakdown of the Human Genome

?% Non-Coding RNAs

1.5% Vertebrate genes

9% HERV & LTR

52.5% Unknown

21% LINEs

13% SINEs

3% DNA Transposons

Breakdown of the Human Genome 2012

How truly extraordinary is the complexity of this structure that lies at the very core of our being! These different genetic entities are not parcelled up neatly in different parts of the genome as we see in the pie chart, they are all jumbled up – virus and vertebrate gene, the DNA that translates to long non-coding RNA, ignoring the supposed function of other coding stretches and cutting into it or straddling several in one stretch. The motley assemblage sits cheek by jowl or piled on top of one another throughout the chromosomes. And hidden within this remarkable and messy reservoir of the heredity of each and every one of us is a secret narrative of our human history, from the most distant ancestors from long before the human stage of our existence right up to the immediate present.

It is to this new mystery I would now like to direct our journey.

fourteen

Our History Preserved in our DNA

This science appeals to us very differently from physics. It directly informs our understanding of ourselves. Its mysteries once deemed dangerous and forbidden: its consequences promise to be practical, personal, urgent.

HORACE FREELAND JUDSON,
THE EIGHTH DAY OF CREATION

On 13 February 2014 the journal *Nature* published an article with the title, 'The genome of a Late Pleistocene human from a Clovis burial site in western Montana'. The Clovis culture is a prehistoric American culture named after distinct stone tools found at sites near Clovis, New Mexico, in the 1920s and 1930s. Dating back to the end of the last Ice Age, roughly 13,000 to 12,600 years ago, the Clovis people are believed by many American palaeontologists to be the ancestors of the Native Americans of North and South America. At the time of publication the origins of the Clovis people were still being debated, with most anthropologists believing they came from Asia although some proposed an alternative route from southwestern Europe, following the margins of the ice sheets across the Atlantic Ocean. The Montana burial site was already of historic importance. First discovered in 1968 on land owned by the Anzick family at the foothills of the Rocky Mountains near Wilsall, it contained the skull and other skeletal remains of a male infant, roughly 12–18 months old, now known as Anzick-1. It was prized as the only known Clovis burial that also included a considerable assemblage of stone tools and bone tool fragments.

The child's remains had already been carbon dated to roughly 12,700 years old – the oldest known burial to date in North America. This, together with the characteristic tools, suggested that he belonged to the earliest phase of Clovis immigration. So the sequencing of his genome might provide invaluable information on the ethnic and geographic origins of the earliest Native Americans. The sequencing work was undertaken by a team of Danish evolutionary biologists together with experts based at the National History Museum and University of Copenhagen.

What then were the scientists really looking for in the genome of this child, who had died during the last great Ice Age?

The answer is that they hoped to learn something about our human origins and migrations back in a time when all of humanity was still dependent on hunting and gathering for survival, when the only tools and weapons were made of wood, bone and stone – a time when there were no national boundaries, no empires, no cities, no agriculture.

To get a clearer grasp of what these palaeogeneticists were looking for we need to understand what they mean by SNPs – an acronym for 'single nucleotide polymorphisms'. It sounds complicated but, as we shall see, it is simplicity itself – if we hop aboard that now-familiar magical steam train to make a new exploration along the railway track of DNA. In this case we choose to steam along a stretch of DNA in a germ cell during the formation of the sperm or the ovum, when I suggest we take a closer look at what sometimes happens when DNA replicates. I hardly need to remind you that the sleepers are made up of complementary nucleotides, C always joining in the middle, through that weak cement of the hydrogen bond, with G, A with T, and vice versa. Now, as we watch the cycle of replication take place, the two sides of the rail separate at the hydrogen bond linkages in the sleepers and the long strands disentangle to begin the process of copying. At my suggestion, we follow this happening along the lowermost rail – the so-called 'anti-sense' strand. We now head eastwards, over thousands of sleepers before I stop the engine. We get down off the train to take a closer look at a single sleeper.

'I should explain that this is a section of DNA in a so-called non-coding part of the genome. So it is not part of a protein-coding gene.'

'What are we looking for?'

'A mistake in the copying.'

Just as when we looked for mutations in protein-coding genes earlier, you spot the error. Where a G in the incomplete sleeper should have attracted its complementary C before the sleepers rejoined to form the complete track, there has been a mistake. A thymine, or T, has taken the place that should have a cytosine, or C. This is another point mutation. Clearly this one nucleotide will not quite fit the G, so the sleeper is buckled. But in subsequent replication cycles, the T mistake will now attract a matching adenine, or A, in copying to a new sense strand. This change in the DNA sequence will be passed on to the germ cells, to be inherited by the offspring of the individual, and so on to future generations. This is what is referred to as a single nucleotide polymorphism – or 'SNP', nicknamed a 'Snip'.

Since the mutation is in a non-coding sequence, it won't affect the health of the offspring. Snips like this are ignored by natural selection – or to use the jargon, they are 'selectively' neutral. The fact that they are selectively neutral means that they are inherited without bias or favour throughout all future generations. Over time, more and more of these Snips accumulate within an inter-breeding species population, creating genetic 'markers' in specific places within chromosomes that identify that particular genetic lineage from that time onwards.

There are millions of Snips in every human genome. And they show significant variation from one individual to another, and a great deal more variation between different human populations. Particular Snips gather as identifiable clusters in a specific region of a chromosome, where they tend to be inherited together as a group, or 'haplotype', even remaining undisturbed during the swapping of bits of matching chromosomes during the sexual recombination that takes place during the formation of the germ cells. I should also explain, in passing, that the original definition of haplotype referred to clusters of *genes* that tended to be inher-ited as a closely linked collection, but this definition had to be modified when we discovered that most of the human genome is not actually made up of genes. If you are male, your Y-chromosome haplotype should be the same as your father's, and to the gener-ations of males going back through time in your paternal lineage. The same rules would apply to mitochondrial haplotypes and maternal lineages, though both males and females inherit their mitochondrial lineages exclusively through the maternal line.

Geneticists also employ another grouping, called a 'haplogroup', which some use to group haplotypes into more distant common groups that share an overall common ancestor. However, I should voice caution here, because some geneticists ignore this distinc-tion, using 'haplotype' or 'haplogroup' to mean much the same thing. For example, Celtic males, such as the original Irish, Welsh and Basques, share a Y-chromosome haplogroup, as do males of Germano-Nordic origins. But if we go further back, most European males, or females, coalesce into a common haplogroup of still earlier origins when compared with, say, males or females of east Asian origins. Thus haplotypes tend to be used for more closely related family trees and haplogroups for more distant historic and archaeological population genetic studies.

A haplogroup (or haplotype) begins with a 'root' or 'founder' mutation, which has been located by a combination of archaeological and palaeogenetic studies to a specific historic human population. This is then added to by subsidiary selectively neutral mutations within the same region of the chromosome, creating recognisable genetic subgroups over time. The root or founder mutations are usually given a capital letter and the subsequent mutations, which come about through additional Snips, are given numbers or lowercase letters. Thus the lineages appear like the branches of a tree – it begins with a trunk, then major branches, and then finer and finer branches, representing different subgroups radiating from the founder group over thousands or tens of thousands, or even hundreds of thousands of years.

One such ancient haplogroup, found exclusively in mitochondrial DNA, is called the 'D' root or 'clade'. This originated as a founder Snip in a population living in northeast Asia, including present-day Siberia, roughly 48,000 years ago. Over time additional Snips arose in the mitochondrial DNA of the descendants of the D population, leading to four divergent clades or branches, called D1 to D4, and additional mutations within the still migrating clades gave rise to further sub-branching over time. Each new branch, or sub-branch, would correspond to a geographic location and timings of population movement, which can be cross-referenced to archaeological findings, such as carbon 14 dating, so population geneticists can plot the historic movements and interactions of people within these clades all over Asia, Europe and in due course to North and South America.

Returning to the child, Anzick-1, we recall that the skeleton was carbon dated to 13,000 to 12,600 years ago, which places him to very early in the human colonisation of the Americas. His mitochondrial haplogroup was found to be D4h3a, which is one of the rare lineages specific to Native Americans. Given the date and the haplogroup the researchers concluded that Anzick-1 belonged to an ethnic group that must be close to the founder of the D4h3a sub-lineage and thus his people were thought to be directly ancestral to 80 per cent of all Native American peoples, and close cousins to the remaining 20 per cent. The study of Anzick-1's genome also turned up some distant commonalities with European haplotypes.

In the same journal, a paper by the same group of genetic and archaeological investigators described a young boy's remains, dating back 24,000 years, from an upper Palaeolithic burial site in Siberia. The oldest burial of any modern human discovered to date, study of his haplotype revealed that he belonged to an even older mito-chondrial haplogroup than Anzick-1 – a basal lineage of haplogroup R. Today this lineage is found in people living in western Eurasia, south Asia and the Altai region of southern Siberia. Sister lineages of haplogroup R form a haplogroup Q, which is the commonest haplogroup in Native Americans, and in Eurasia the Q haplogroup lineages closest to those of the Native Americans are also found in the Altai region of southern Siberia. In the opinion of Danish palaeontologist Eske Willerslev, who led the sequencing of both sets of remains, 'At some point in the past a branch of east Asians and a branch of western Eurasians met each other and they widely interbred.' Their descendants headed east, across the land bridge between Asia and North America, discovering two huge and boun-teous continents that had never been populated by humans before. They gave rise to the majority of Native Americans we see today, including Anzick-1. While not everybody agrees with Willerslev, the combination of the two infant boy discoveries would explain how Native Americans share 14 to 38 per cent of their genomes with western Eurasians.

*

Snips, haplotypes and haplogroups are not exclusive to our nuclear genome. I referred to mitochondrial DNA in both these genetic explorations and now we can hop back aboard our train for another trip into that mysterious ultramicroscopic world to probe yet another mystery. But our destination on this trip is no longer the landscape of the nuclear genome; this time we are heading into the territory that lies outside the nuclear membrane, into the equally intriguing landscape of the cytoplasm, moving carefully through an incredibly congested and frenetically busy space which might be compared to an industrial landscape, in which fresh proteins are being manufactured, ageing proteins being broken down for recycling, and where huge machines, resembling free-floating sausage-shaped juggernauts, are extracting energy from gaseous oxygen and packaging it in storable form, so it can be used in every living cell. These are our vitally important human mitochondria.

As we watch, a mitochondrion develops a constriction about its centre, and before our startled gazes it buds, the parent mito-chondrion cleaving itself into two clones. As our steam engine carries us deeper into the ultramicroscopic world of the genes, we find ourselves passing through the outer wall of one of the sausage-shaped juggernauts, which, as we magically contract in size, has now grown comparatively gigantic, fully to the size of a city. Within its cavernous spaces we come upon another railway track, with its gleaming rails of deoxyribose sugar and stiffening phosphates, and the familiar sleepers, with their interlocking nucleotide bases, heading away into the blurry distance. We have entered the world of the mitochondrial genome, with its very different evolutionary origin from that of the nuclear chromosomes, an enigma within the greater mystery.

'I know you talked about mitochondria – about where they came from . . .'

'A genetic union between what was once a free-living parasitic bacterium and the single-celled forerunner of all complex life on Earth. The mitochondria still retain quite a lot of their bacterial origins. They retain enough of their genome to reproduce by themselves, which is what that mitochondri was doing. While each of us inherits half of our nuclear genome from our father, we inherit more than that from our mother. In addition to half her nuclear genome, we also inherit the physical structure of a cell – the ovum – which includes the mitochondria.'

'Which is why we all, males and females, get our mitochondrial genetics only from our mothers?'

'Yes. And that explains why our mitochondrial inheritance doesn't follow the Mendelian laws of nuclear inheritance, such as recessive and dominant behaviour of genes. The mitochondria also reproduce much more frequently than the nuclear genome, and because they are bacterial genes, and thus less able to correct mistakes, they are even more liable to mutation.'

'So – that's how you get mitochondrial Snips? Haplotypes – and haplogroups?'

'You've got it! And these will be exclusively inherited through our mothers so they will run true to the maternal line, right back to the year dot.'

'So there was a woman, back in Elizabethan times, who would have had the same mitochondrial DNA as my mother?'

'The same as you – other than whatever Snips have accumulated in the meantime. And you can go much further back than that. Why not to Roman times, or the beginnings of agriculture in the

Fertile Crescent? In fact, you could go much further back still, to the ovaries of the maternal line to the very origins of *Homo sapiens.*'

As we tootle our whistle and steam along this new railway of mitochondrial DNA, I would like to explain a little more about the way our human history is written into our genome – or, if you like, our two symbiotically linked genomes. One of the key things to grasp is that we have, in essence, three different parts of our holobiontic genome that are libraries of three different genetic histories. One is the mitochondrial genome, which tells us the story of the maternal genetic line – the matrilineage. The Y chromosome, which is part of the nuclear genome, tells us the story of the paternal genetic line – the patrilineage. And the remaining nuclear lineage, which is by far the bulk of our genetic inheritance, tells us far more about our genetic history as a species.

All the while we've been rattling along this mitochondrial railway, we have been glimpsing subtle differences in how this bacterial genome works. For example, the mitochondrial genome is much smaller than that of a single chromosome within the nucleus. The total mitochondrial genome consists of 16,600 nucleotide pairs – sleepers – where the nuclear genome, even in the halved genome of the germ cells, consists of 3.2 billion. There is no shake-up of the mitochondrial chromosomes, as we find during sexual reproduction involving the nuclear genome. In fact, the mitochondrial DNA isn't composed of linear chromosomes at all. The mitochondrial railway consists of a single circular track, as we would find in a bacterial genome, so that, if we take a lengthy journey along the mitochondrial railway track, it will eventually bring us right back to where we began.

'This symbiosis – the event, as you call it, that gave rise to mitochondria – you said it only ever happened once?'

'We know this for certain because all the mitochondria, from every animal, plant, fungus and the oxygen-breathing less complex organisms, are clearly derived from a single ancestor.'

'You can determine this from the mitochondrial genes?'

'Yes.'

'Why then, if this happened so very long ago, have the two genomes, the mitochondrial and nuclear, not joined up? Wouldn't that have made sense?'

'You're right. It would have made sense. In fact, most of the structural proteins in what we now call the mitochondrial organelles are coded by nuclear-based genes. We believe that at least 300 of the nuclear genes were formerly mitochondrial genes that transferred.'

'But some didn't?'

'Those genes that stayed within the mitochondria are all, or nearly all, involved in the respiration of oxygen. Oxygen is an extremely toxic element. It's possible that it had to be handled within separately walled-off organelles to prevent its toxicity affecting the remainder of the living cell.'

'But wasn't oxygen always around, as part of the atmosphere?'

'No. Atmospheric oxygen is produced by plants and cyanobacteria. It is a by-product of their internal chemistry. We might recall that the early arrival of oxygen into Earth's atmosphere proved calamitous to many of the oceanic and shore-dwelling life forms at that time. Only those who could breathe oxygen could survive its toxic presence. And even within the cells of those who inherited the mitochondrial ability to breathe oxygen, it had to be contained away from the delicate genomic machinery within the nucleus. It had to be locked away inside the original invading cells, the former bacteria that had now evolved into the mitochondrial organelles that were already resistant to the toxic effects of the oxygen.'

'Didn't we see something very similar with the viruses within the chromosomes?'

'We did! Those viruses that were expressing their genes as proteins appeared to retain their original genomic structure, including the controlling promoters.'

But to return to the use of mitochondrial genetics in palaeontology, the uniquely maternal lineage of mitochondria has inevitably provided a powerful tool for evolutionary geneticists in exploring population genetics and the complex weave of human movement throughout history. Thus we are not surprised to discover that the D4h3a haplotype component that linked the Clovis child to his Siberian ancestors came from the study of Anzick-1's mitochondrial DNA.

As mentioned above, another equally powerful tool for population geneticists comes from the Y chromosome within the nuclear genome of males. Like the mitochondrial DNA and its exclusive link to the maternal genetic line, the Y chromosome is exclusively passed down from fathers to sons. Since it has no matching chromosome to recombine with during sexual recombination, it doesn't undergo the jumbling of bits from one chromosome to another. This provides a very useful means of following the pattern of key mutational haplotypes in different Y-chromosome populations over time. For example, in the case of the Clovis infant, Anzick-1, the 'founding Q haplotype' on his Y chromosome was found to belong to subgroup L54*, which the geneticists predicted had separated out from another founding haplotype subgroup, M3, dating to approximately 16,900 years ago. This confirmed that Anzick-1 belonged to an ethnic population grouping that was closely related to the first humans to arrive into the Americas.

*

Just how accurate is this exploration of haplotypes and haplogroups? King Richard III of England was made infamous by the Shakespeare play that bears his name. The last of the Plantagenet dynasty, Richard was killed in the Battle of Bosworth Field on 22 August 1485, during the bloody War of the Roses. Shakespeare's play portrays Richard as a hunchbacked villain, who murdered his brother and two young nephews before attempting to marry his niece. But his supporters, including Philippa Langley of the Richard III Society, defended his reputation, claiming that Shakespeare had defamed Richard in support of the Tudor monarchs who had supplanted him and who reigned at the time when Shakespeare wrote his play.

In an exquisite story of detective sleuthing, Langley traced the historical records to find that Richard's remains had been buried, without coffin or dignity, in an Augustinian Friary in the city of Leicester. Having obtained a small amount of funding she persuaded the archaeologists of the University of Leicester to conduct a dig in what was a present-day car park which was thought to overlie the former Friary high altar. There they discovered a human skeleton which fitted the historical descriptions of Richard. It was that of a man in his thirties with a severe curvature in his thoracic spine, known as a scoliosis – fitting Shakespeare's description of Richard's hunched back. It also showed signs of multiple wounds, indicating that the man had died in battle. Radiocarbon from two independent sources dated the bones to between 1430–1460 and 1412–1449. This seemed too early for Richard, but mass spectrometry carried out on the bones suggested that their owner had eaten a good deal of seafood, which can skew radiocarbon dating. A corrected date worked out at between 1475 and 1530, which would fit nicely with the historic date of Richard's death. However, uncertainties remained and, even with the location, the physical anatomy and the radiocarbon dating, controversy still reigned as to whether Richard's remains had been found.

A new line of genealogical research managed to identify a

woman, Joy Ibsen (née Brown), who was a direct matrilineal descendant from Richard's mother. In genetic terms, Mrs Ibsen should be carrying the same mitochondrial haplotype as Richard himself, but she had emigrated to Canada after the Second World War and died there in 2008. Fortunately she had given birth to a son, Michael, who was prepared to give a sample of his DNA for testing. The controversy was resolved when Michael Ibsen was found to share the rare mitochondrial haplotype J1c2c with the exhumed skeleton. There could no longer be any reasonable doubt that these were the bones of the last Plantagenet king.

As this genetic exploration of our human inheritance continues, we have become increasingly aware of how accurately our genome reflects hugely important movements and evolutionary events in our human ancestry and history. If some researchers are to be believed, we may even have discovered the genetic Adam and Eve.

fifteen
Our More Distant Ancestors

Our obsession with fossils has distracted us from a much richer source of evolutionary information: genetic data . . .

LUIGI LUCA CAVALLI-SFORZA

We are all more closely related to one another than we might imagine, as a little thought experiment will quickly demonstrate. If we were to consider, say, four new generations of our ancestors per century, we can readily construct a branching tree with the numbers of ancestors doubling with each generation as we travel backwards. Four grandparents descended from eight great-grandparents, who in turn descended from 16 great-great grandparents, and so on. Two centuries ago, or eight generations, we discover that we descended from some 256 multi-great ancestors living in that single generation. In four centuries the number rises to 65,536. In eight centuries the number rises to 4,294,967.296, – which is vastly more than the entire population of the world at that time – probably more people than had ever lived on Earth up to then. We need hardly go further back to realise that there is something seriously amiss with this line of thinking. Since we could not possibly have so many ancestors there has to be another explanation.

The answer is simple: we all have a great many common ancestors. This can be explored at genetic level by constructing haplotype trees, and it can be extended even wider geographically, and deeper into the past, if we construct haplogroup trees. The closer people are to one another, the more haplotypes and haplogroups they will have in common. And each distinguishing haplogroup is a marker – a genetic signpost – to a single ancestor in a specific place and a specific time who was the first to inherit the relevant Snip. From this haplogroup marker specific population groups can be identified and their subsequent movements and migrations plotted.

The Italian geneticist Luigi Cavalli-Sforza, who was a professor at Stanford University, devoted his life to gathering this type of genetic information on different human populations. In his book *Genes, Peoples and Languages*, Cavalli-Sforza dismantled the idea of different races, arguing that the differences we saw between Africans, Asians, Europeans and Australasians, were superficial evolutionary adaptations to local conditions such as climate. Genetic studies of different peoples throughout the world have confirmed that we are all part of a single species, in which our commonalities overwhelm our differences. Archaeological studies of fossilised bones, commonalities of tools, and the patterns of habitation and culture all point to the likelihood that our human ancestors originated in Africa, most likely sub-Saharan East Africa. These traditional archaeological disciplines are now reaffirmed and augmented by genetic studies that allow us to gather much deeper perspectives on our human history than was possible before.

We have already glimpsed how specific clusters of mutations, known as Snips, in the mitochondrial genome and on the Y chromosome, enable geneticists to trace genetic haplotype and haplogroup lineages, and thus chart human population migrations, back into prehistoric times. Similar haplotypes and haplogroup lineages can be found in the 22 pairs of human chromosomes unallied to sexual differences, the so-called 'autosomes', enabling a third line of genetic tracing of human lineages and population movements. The distribution of specific endogenous retroviruses can be treated in the same way to locate the African origins of *Homo sapiens* and the subsequent global migrations of our species, which at this early stage are called 'early modern humans'. The genomic viruses also have a role to play in such genetic determinations. For example, the distribution of two human-specific endogenous retroviruses, HERV-K113 and HERV-K115, are adding to the history.

Unlike most of the other HERVs, these two appear to have entered the genome after the primary migration of early modern humans out of Africa. Thus, while the great majority of HERVs are common to all of us, these two are found in a high percentage of present-day people who hail from East Africa, Arabia and further east into Asia, but in low percentages, or not at all, in people hailing from Europe. To some geneticists this suggests that there may have been more than one migration of modern humans out of Africa, perhaps with expansions and retreats that may have been precipitated by significant environmental or climate change.

As we have seen, mitochondrial genetics offers a series of mutational tags that make it possible to follow some of these complex population movements. Without the influence of evolutionary change, every daughter will inherit exactly the same mitochondrial genome from her mother, again and again, throughout all of history. If this were the case, my mitochondrial genome, inherited from my mother, would be identical to that of a common ancestor in Africa, say, 200,000 years ago. But we have seen how the mitochondrial genome has been altered by copying mistakes, or 'mutations', during the budding style of reproduction that mitochondria undergo. Sometimes these copying mistakes cause impairment of function of the mitochondrial genome, which would have very likely given rise to disease. But such pathological mutations would not become established as lineage markers because the resulting disease would result in reduced reproductive fitness. Only mutations that had no significant effect on reproductive fitness would have become incorporated as lineage markers. These have become part of haplotypes and haplogroups, which, having no effect on survival, are ignored by natural selection, so they survive unchanged over vast time periods.

Snips like this crop up at reasonably predictable intervals – a feature that enables geneticists to compare the numbers of mutations in a given stretch of the genome to a 'molecular clock'. We have seen how key clusters of mutations in certain regions of the mitochondrial genome – haplotypes and haplogroups – can be linked to founder individuals, in place and time, then spread into the descendant population, allowing the migrations and movements of this population to be plotted in geography and time. Looked at from a different angle, differences between haplogroups are markers of different historic populations. And where we find, albeit rarely, a single haplogroup that is common to very many widely dispersed populations, this is seen as an important marker linking all these populations to a common founder ancestor.

What then if we were to discover a mitochondrial haplogroup that is common to every man and woman on Earth today? Would this not point to a woman who was a common ancestor of all of us – a genetic Eve?

In the 1980s this was what a group of geneticists, led by Allan Wilson, of the University of California, Berkeley, had in mind when, with the help of his doctoral students Mark Stoneking and Rebecca L. Cann, he conducted an examination of mitochondrial DNA from 147 Americans coming from a wide variety of racial and ethnic groupings. They were looking for evidence of shared and divergent haplogroups that would enable them to construct a hereditary tree for all of humanity. A decade earlier, Wilson had been joined by another pioneering geneticist, Wesley M. Brown, who had developed new techniques for screening mitochondrial DNA. Between them these scientists discovered that mutation of mitochondrial DNA was 5–10 per cent faster than in nuclear DNA. It was Wilson who first thought of the idea of the molecular clock, based on the fairly predictable occurrence of mutations in the human genome with time. Now they were convinced that they had the tools to investigate the potential of mitochondrial mutations as a measure of evolutionary relationships over time.

In the paper, Wilson and his colleagues figured that the global human population broke down into two broad mitochondrial haplogroups. One of these was confined to Africa; the other included some African groups as well as the rest of humanity. They drew several conclusions. For a start, it appeared to confirm the 'out of Africa' theory, proposed by some palaeoanthropologists for the origins of *Homo sapiens*, while contradicting the alternative 'multi-regional theory' which proposed that modern humans had not come directly out of Africa but had evolved over vast time periods in the major continents. It also supported what is now known as the 'recent common origins' of modern humans, which proposes that all of the people on Earth today are part of a single, closely related population that emerged from Africa some time in the last 200,000 years. Their extrapolations went further; the African haplogroup had the greatest genetic diversity, a finding that has since been amply confirmed by other studies. There can be more genetic diversity, as defined by Snips, between neighbouring African peoples, for example, across a major river, than we find across the entire Eurasian landmass. If we then consider that Snips arise through mutations at a fairly predictable rate over time, this implies that *Homo sapiens* has lived in Africa for far longer than anywhere else on Earth.

However, there was an additional, altogether surprising extrap-olation. Wilson and colleagues also claimed that they had found genetic evidence for a female 'last common ancestor' of modern humans – a woman in Africa who was the first to acquire the founder mitochondrial haplogroup mutation that is common to all of the people on Earth. They figured that this woman, dubbed by the media as 'mitochondrial Eve', contributed the common founder haplogroup to the human pedigree some time between 140,000 and 200,000 years ago. The idea of a mitochondrial Eve created front-page news headlines, proving both exciting and controversial at the time. It was, perhaps inevitably, misunderstood by many lay people, including religious groups, to imply that mitochondrial Eve was the single female ancestor of all of us.

We should maintain a prudent caution here – for reasons that will become clear a little later. For the moment, let us celebrate our common matrilineal ancestor with a visit to meet her in her hunter-gatherer community, and look at how she has been linked to the entire world.

We can reasonably assume that she would have appeared little different to the other women in her small, close-knit community. Some believe that she would have resembled the San people, who, until very recently, still followed a hunter-gatherer existence in southern Africa, with the men spearing fish in the shallows or hunting land animals for meat, the women digging for roots or foraging along the shoreline for shellfish. We know that she had the gift of language, with all of the social potential that conveys. We also know that she might have painted her skin or clothing with patterns in ochre. We can hazard a guess at what might have been her clothing – perhaps a covering skirt from the waist down made from plant fibres, or animal hides. A study by the University of Florida found evidence that humans may have begun to wear clothes as early as 170,000 years ago. She is also likely to have decorated her neck, wrists, or clothing with beads of small, similarly sized sea shells, which were brightly coloured with natural pigments. We can also pretty much assume that the older females taught the children, and younger adult females, what to do, in foraging through forest and seashore, and in the bearing and caring for children.

Mitochondrial Eve was not the ultimate great-grandmother of all humanity. She would have been one of many females alive and reproductively active at the time she acquired the founder haplogroup. All of these other reproductive women would have been just as likely to contribute to the species gene pool, but she was the only one whose mitochondrial genome found its way into all modern humans. Let me explain how this is likely to have arisen.

Let us say that there are ten reproductively active women in a single hunter-gatherer group. Only one of these, Eve, has the founder mutation, or small cluster of mutations, in her mitochondrial genome. Perhaps eight of the women are reproductively successful, giving rise to two or three surviving offspring per woman. Eve, for example, might have given birth to two surviving daughters and a son. All three will inherit her mitochondrial genome, but her son will not pass it on – males do not contribute to the mitochondrial lineage. Other women in the same group may, through chance, have had no daughters, so they will not have contributed to the mitochondrial lineage. In subsequent generations the same actions of chance will continue to operate. Generation after generation, for some 140,000 to 200,000 years, Eve's descendants must have given birth to a daughter in every generation for this to result in an unbroken matrilineal lineage over the vast time period to the present day.

So now we understand how really it was a game of chance that resulted in mitochondrial Eve drawing the winning ticket. But this does not mean that those other mothers – and fathers for that matter – have not contributed to us all genetically. We have already seen how, using simple mathematics, we share very many common ancestors, who will have contributed to other aspects of our genome.

Mitochondrial Eve would have almost certainly lived in Africa, although where in Africa is uncertain; perhaps somewhere in the region of modern-day Tanzania. Eve's founder mitochondrial haplogroup is the 'macrohaplogroup' L and it probably did originate somewhere between 120,000 to 200,000 years ago. Current thinking suggests that her matrilineal line first spread out over the remainder of Africa, the original L haplogroup evolving to regional subgroups L0 and L1 to L6. Although we tend to imagine these ancestral populations radiating out from East Africa into the Middle East, and from there radiating west to Europe and northeast and southeast to Asia, Australasia and the Americas, in fact the genetic evidence suggests a complex admixture and movement with waves of advance and return before, perhaps, a new migration, starting about 60,000 years ago, saw the mitochondrial haplogroup L3 first diversify into haplogroups M and N in East Africa before crossing into the Arabian Peninsula, from where it spread and diversified, perhaps through a coastal migration, into Asia, Eurasia, Europe and the New World. This would imply that all the mitochondrial lineages outside of Africa descended from the M and N lineages.

These basal M and N lineages have now been traced along the southern Asian shoreline. A combination of archaeological and genetic evidence has also revealed that as the expansion and migration progressed over thousands of years, these M and N subgroups acquired further defining sub-subgroups, as the populations moved, and the lineages split into smaller and smaller branches. For example, if we screen modern populations we can infer that those with mitochondrial haplogroups H, I, J, N1b, T, U, V and W are of European origin and those with A, B, C and

D are of Asian and the New World origins, with G, Y and Z predominantly associated with western Asia.

*

It is in the nature of scientists to be sceptical, and the scientist in me asks the question: Are we being overly simplistic in assuming that this sequence of mitochondrial haplogroups extrapolates to actual population movements and present human diversity?

Two geneticists, Brigitte Pakendorf and Mark Stoneking (the latter one of the original Wilson group at Berkeley), have warned us that the studying of mitochondrial haplogroups has limitations when it is extrapolated to explain major population movements. They don't deny that it is a useful tool, but they advocate expanding the searches to the analysis of the entire mitochondrial genome and further, to a much wider genetic analysis. A very obvious next step would be to extend the genetic analysis to the patrilineal ancestry. So where in all this historic exploration is the genetic evidence for the ancestral Adam?

Just as the mitochondrial inheritance passes entirely through the maternal genetic lineage, we assume that the Y-chromosome nuclear inheritance passes entirely through the paternal genetic lineage. This is based on the Y chromosome having no corresponding partner to recombine with during the formation of the sperm cells, so it is subjected to no mixing of chromosomal elements from both parents during the formation of the germ cells. In fact, this is not completely true. Some 5 per cent of the Y can and does recombine with a corresponding part of the X chromosome during germ cell formation. But geneticists get around this by focusing on the 95 per cent that is invariably passed from father to son. This is called the 'male specific region' of the Y chromosome, or 'MSRY'.

Acronyms – brrrh!

Perhaps, like me, you have an instinctive aversion to acronyms? Alas, we need to get the hang of yet another two that geneticists are fond of flinging into the heated air of debate when it comes to matrilineal and patrilineal lineages – namely LCA and MRCA, which in fact denote exactly the same thing: the 'last common ancestor' and the 'most recent common ancestor'.

Brrh – and brrh again!

In contrast to the mitochondrial genome, which comprises roughly 16,000 base pairs, the Y chromosome comprises 60 million. This means that the study of Y-chromosome mutations, those Snips and haplotypes and haplogroups, is more complex than that of mitochondrial mutations. But the consolation is that since they focus on two different genomes, with different evolutionary and genetic origins and thus different mutational rates and properties, the sum effect of combining the two adds significantly to the accuracy of all those archaeogenetic calculations.

For a start, studies of Y-chromosome haplogroups also point to Africa as the place where modern humans evolved. But in this case it suggests either eastern or southern Africa as the place where 'Adam', the earliest detectable common male ancestor, or 'Y-MRCA', was born.

A basal haplogroup lineage, conveniently labelled Haplogroup A, is more frequently found in males from Africa than anywhere else in the world. Adam's arrival on the scene was variously estimated as 188,000 years ago – or 270,000, or 306,000, or 142,000 or 338,000 years ago. Some of this disagreement may have come about from different ways of calculating the so-called 'molecular clock', but it may also have resulted from problems with the genetic analysis of very long DNA sequences. More recently a study by G. David Poznik and colleagues in 69 males from nine very different populations, and one by Paolo Francalacci and colleagues from 1,204 Sardinian males, arrived at estimates of the

Y-MRCA, the mean of which tallied a little more closely with the purported age of our common maternal ancestor. These offered a range of from 120,000 to 300,000 years ago.

Y-chromosome Adam clearly did not inhabit an African Eden at the same time as mitochondrial Eve. Nevertheless, through such genetic studies the evidence that modern humans originated in Africa gathered momentum. But where we might have looked for helpful confirmation from the fossil record over the same key period of 100,000 to 300,000 years, this proved somewhat elusive. Chris Stringer, at the Natural History Museum in London, drew attention to this prevailing lack of palaeoanthropological information in an article in the journal *Nature* in 2003. But he was pleased to draw attention to two reports in the same issue of the magazine, which described three fossilised skulls from close to the village of Herto in Ethiopia that were, in his opinion, some of the most significant palaeontological discoveries of early *Homo sapiens* to date.

The skulls, which included adults and one juvenile, were almost complete and their antiquity, at about 160,000 years old, tallied remarkably well with the genetic dating. They were associated with stone tools of the so-called Acheulean, or Middle Stone Age, types. The palaeontologists also found evidence that the heads of the adults and juvenile had been removed from the bodies after death, with the lower jaws deliberately disarticulated and the skulls systematically defleshed. While this could have been associated with cannibalism, which is sometimes found in human fossils, there were additional 'more decorative' cut marks, incised by a very sharp and fine blade, that suggested an alternative explanation. These, combined with the polishing of some of the rounded surfaces of the skull, suggested formal mortuary practices followed by cultural or ritualistic treatment. The patterns of the marks were similar to what is seen in the skulls of some much more recent

New Guinea crania, which were known to play a role in ritual mortuary practices.

*

This palaeoanthropological evidence, combined with the historic evidence stored within our human DNA, is providing powerful corroborative evidence for the origins of modern humans in Africa. Yet further mysteries need to be resolved. When did our distant African ancestors move out to populate the rest of the world? Did they do so in a single historic migration? Or, given human migratory behaviour in more historic times, together with the major climatic variability produced by the comings and going of Ice Ages and natural calamities such as volcanic eruptions, is it more likely that there was a series of ebbs and flows, like smaller wavelets overlapping two or more major waves of migrations over time?

The Wilson study mentioned above, and many subsequent studies of human genetic diversity, highlight what may be a related puzzle. If we compare our human genetic diversity to, say, that of our evolutionary cousin, the chimpanzee, we discover significantly less genetic diversity in the human genome. This is particularly noticeable if we compare the important genetic region known as the major histocompatibility complex, or MHC, which, as we saw earlier, determines immunological and biological self and plays a vital role in our immunological reaction to invading infectious organisms, such as viruses and bacteria. This loss of diversity is a significant finding. It suggests that at some time in our evolutionary history – and some geneticists believe that it was a time very close to the origins of early moderns as a species, or perhaps close to the timing of the spread of the species out of Africa – that the ancestral human population was subjected to a near-extinction event. This created a genetic bottleneck that reduced the founder population to less than 10,000 individuals, and some even think it may have been as few as 1,000. What possible disaster could have brought about such a shocking cull?

One suggestion was the explosion of a volcano called Mount Toba on the island of Sumatra, which is believed to have erupted 70,000 years ago. But if this distant catastrophe was capable of slate-wiping populations extending from east Asia to Africa, it would have wiped out the human population closer to the epicentre. The survival of populations in India, as shown by primitive stone tool assemblages in layers above the ash, brings this into question. Moreover, the distinguished palaeontologist Sir Paul Mellars, at the University of Cambridge, has produced convincing evidence that modern humans are more likely to have reached Asia, through a coastal migration, at least 10,000 years after the volcanic explosion, which would cast additional doubts on Mount Toba being the causative catastrophe. There is, however, another contender, one that brings us back to the endogenous retroviruses that account for roughly 9 per cent of our human DNA.

We recall that these endogenous viruses entered the pre-human and human germ line during retroviral epidemics. The most recent of the genomic viral invaders are called the HERV-Ks, a group that first invaded the ancestral primate genome somewhere around 30 million years ago. The evolutionary virologist Luis P. Villarreal, based at the University of California, at Irvine, believes that the arrival, and explosive colonisation, of the primate genome by HERV-Ks was a watershed event in primate and subsequent human evolution. It coincided with the switching off of an earlier viral colonisation of the human genome, by DNA-based viruses or so-called 'transposons'. Most of the HERV-Ks are present, in the same chromosomal distributions, in all of us, many now having entered into useful holobiontic function with the rest of the genome. At least ten subgroups of the HERV-Ks invaded the human germ line after our separation from chimpanzees, so these

HERV-Ks are exclusive to the human genome. Four of these are believed to have entered the human genome during the last million years, including HERV-K106, HERV-K113, HERV-K115 and HERV-K116. Based on the molecular clock – in this case we are applying it to DNA mutations in the regulatory regions of the viruses, known as long terminal repeats, or LTRs – HERV-K115 inserted into human chromosome 8 roughly a million years ago, while HERV-K113 inserted into chromosome 19 roughly 800,000 years ago. HERV-K116, which inserted into chromosome 1, and HERV-K106, which inserted into chromosome 3, have no mutations in their LTR regions. This suggests that their insertions, and the relevant exogenous retroviral epidemics, were much more recent than HERV-K115 and HERV-K113.

In 2011, Jha and colleagues reported the research outcome of six different groups of American genetic and evolutionary scientists who combined forces to examine the distribution of HERV-K106 in 51 Americans of diverse ethnic origins. This allowed them to separate the test populations into four different haplogroups. They concluded from the haplogroup evidence that HERV-K106 inserted into the human genome roughly 91,000 to 154,000 years ago. This must have resulted from a retroviral epidemic affecting the human population at the same time. Yet unlike HERV-K115 and HERV-K113, HERV-K106 was found to be universally present in the genomes of their test subjects, suggesting it was ferociously infectious. Judging from the patterns of the two prevailing retroviral pandemics, HIV-1 in humans and the koala pandemic in Australia, exogenous retroviruses are extremely efficient in their spread as emerging infections within a geographically contiguous species. And we know how they behave in relation to their new hosts. Think of AIDS in a virgin human population with no knowledge of epidemiology and no therapy; for its endogenous version to be universal, the exogenous K106 is likely to have swept through all of the human population that contributed to the descendant people who populate the Earth today. And the date of virus insertion appears to coincide fairly well with the approximate dates drawn from both the mitochondrial and the more recent Y-chromosome calibrations, as well as the fossil discoveries from Middle Awash in Ethiopia. In time, study of ancient human genomes, including a careful examination of the endogenous viral components, may help to answer whether or not HERV-106 was the culprit for the human culling that produced the apparent genetic bottleneck.

*

Putting aside such speculations, the weight of evidence is pointing strongly to modern humans having originated in Africa at a date roughly about 150,000 years ago. But this also leaves us with a new set of questions. When did modern humans migrate from Africa to populate the rest of the world? Was there just a single major migration, or was there more than one? We know that our ancestors encountered other, so-called 'archaic' human species when they entered Europe and Asia, so what happened when they met up with them?

sixteen

The Great Wilderness of Prehistory

... in the second stage we had to go into the great wilderness of prehistory to see whether there were elements of internal consistency which would lead one to believe that the method was sound or not.

WILLIAM F. LIBBY

History fascinates us because it tells us how we, and the society in which we live, came to be. For most of us this tends to be a distinctly parochial fascination, the town or county in which we live or, in the widest sense, the country or continent we feel that we belong to. This is altogether natural since it is the world we are familiar with. But there is a deeper history, older by far than national or even continental boundaries – one that takes us back to a time when life was simpler and yet more challenging. There were no schools, no employers and employees, no farms, no herded animals with their supplies of milk and ready meat, no shops with their implicit exchange in the form of money, and above all no metal tools – no machines. This lost world is what Libby, the

225

pioneer of radiocarbon dating, refers to as the 'great wilderness of prehistory'. It is a deeply fascinating world, and one that we know very little about. It is also a world that constituted an extremely important phase in our human history, one that goes beyond national, ethnic and indeed all divisional boundaries, because it is a world that every single human being on Earth living today has in common. We are all descendants of these early modern humans and thus we have a common interest in how these ancestors came to evolve in Africa, and how, out of Africa, they came to colonise the rest of the planet.

Crucial to understanding this stage in our history is the archaeological exploration of the timing of the first evolutionary emergence of early modern humans, a topic we touched upon in the previous chapter. And from there it is equally crucial to establish a reliable framework, in date and geography, of the movements of these pioneering ancestors during their migration, or migrations, out of Africa. This has proved extremely challenging for archaeologists, in part because of the paucity of archaeological evidence discovered to date, and because most such sites are situated in warmer areas of Africa, Europe and Asia, where preservation of organic remains tends to be poor and therefore not easily dated. But this is now changing; windows of opportunity are opening up through new scientific techniques of dating and the genetic extraction from animal bones and human fossils, as well as through a broadening of geographic horizons. On 1 May 2014, I travelled to Oxford to interview Katerina Douka, of the Oxford Radiocarbon Accelerator Unit, in the hope of getting to know more about these interesting developments.

Dr Douka was born in Greece; after graduating from the University in Athens with a basic degree in archaeology and archaeological sciences she went to Oxford to study for a Masters degree and, ultimately, a PhD. Her topic was the dispersal of modern humans out of Africa and into Europe, and within this broad theme her primary interest was radiocarbon dating. It was one of her published papers that introduced me to Libby's concept of the great wilderness of prehistory, and her interest in radiocarbon dating was one of the keys that might open the lock onto this wilderness. I asked her if she had become interested in archaeology at school – or if perhaps a member of her family had stirred her curiosity.

'No. If you come from Greece you are a little bit overwhelmed with archaeology because it is everywhere. I'm just interested in the past, whether it is two hundred years ago or two hundred thousand years ago.'

Perhaps, I ventured, it was people who really interested her?

'Yes – the past and its relation to people.'

The theme of Dr Douka's PhD thesis was: 'The dispersal of modern humans out of Africa and into Europe – from a radiocarbon-dating perspective.'

In 2006 Sir Paul Mellars produced what is now a famous paper in which he described how modern humans were likely to have entered Europe. If Mellars is right, a series of breakthroughs in technology took place in Africa roughly 60,000 to 80,000 years ago. Powerful evidence for this was found in layers of the Blombos Cave in South Africa, which dated to roughly 75,000 to 55,000 years ago. The advances included new patterns of fine stone blade technology, scrapers for working skins and hides, tools for the shaping of bone and wood, bone points for the tips of throwing spears and sharply pointed awls, and carefully shaped artefacts for throwing spears and even, conceivably, the first arrows. These were found in the same sediments as perforated shells used for personal ornament, the earliest such items ever found, as well as large quantities of imported red ochre, including pieces incised with geometric ornamentation. All of this was accompanied by evidence of large-scale exchange and distribution over distances. Equally significant was the evidence, gathered by other scientists, of an accompanying rapid population growth in the ancestral African population, between 60,000 to 80,000 years ago.

These specific African tools and cultural ornamentation show a striking resemblance to what was appearing in archaeological sites throughout the Middle East, Europe and Asia, taken as evidence for the migration of modern humans out of Africa and into these landscapes, roughly 45,000 to 50,000 years ago. It was, of course, this very expansion of the ancestors of modern humans into Eurasia that had been the focus of Katerina Douka's PhD and it has further extended to her present field of scientific interest.

One of her recent scientific papers, which bore the title 'Exploring "the great wilderness of prehistory"', described how the transition from a Neanderthal-dominated western Eurasia to a continent in which modern humans were now the sole inhabitants marked one of the biggest transformations ever witnessed in this vast region. I was aware that some scientists believed that there had been at least two major migrations of early modern humans out of Africa: one about 120,000 years ago, related to some hominin fossil discoveries in the Skhul and Qafzeh cave sites, then in Palestine, and another migration, dated to roughly 45,000 to 39,000 years ago, linked to a famous rock shelter site in Lebanon, known as the Ksar Akil which was also associated with hominin fossils. The latter site was widely regarded as a key staging post in the most important migration, starting in or passing through the Near East. To this end, Katerina Douka led a team that set out to apply modern radiocarbon dating to the human settlements at Ksar Akil.

I asked her about Libby's intriguing phrase: the wilderness of prehistory. What had he meant by it? Did he mean that prehistory in itself was a wilderness in the sense that there was very little known about it?

'Libby is one of the greatest, if not the greatest, figure in radiocarbon dating. He developed the ideas that led to carbon dating while working as a physicist on the Manhattan Project, and for these ideas he later received the Nobel Prize in Chemistry.'

Radiocarbon dating is based on the known rate of decay of the carbon isotope Carbon^{14}. Basically, the atmosphere contains different isotopes of carbon in the form of carbon dioxide, which include the main stable isotope, C^{12}, as well as the unstable isotope, C^{14}, which are present in roughly constant proportions. These isotopes are taken up by plants and micro-organisms, from where they enter the living bodies of other life forms. Once an organism dies, there is no possibility of replenishing the carbon isotopes, and from that moment the ratio of C^{14} to C^{12} will progressively fall over time as the unstable isotope decays to the stable one. Measurement of the ratio of these isotopes is the basis of radiocarbon dating, which has revolutionised the dating of Palaeolithic archaeology sites. This, I presumed, was Libby's contribution to exploring the wilderness.

'That's right. If we look at scientific papers pre 1960, there is no absolute dating. People are talking about a few thousand years, or tens of thousands, or hundreds of thousands, but there was no way of quantifying this as far as time was concerned.'

'How far back can you accurately go with radiocarbon dating?'

'Fifty thousand years.'

'But there's a big chunk of time before that?'

'That's our actual limitation. But in fact we are lucky because the upper limit of this method allows us to calibrate the last years of the Neanderthals and their interaction with modern humans in Europe and more broadly Eurasia. For previous epochs we need to use other methods, such as thermoluminescence, which can take us back 200,000 or 300,000, maybe even half a million years.'

I was interested to know more of the work she did for her PhD.

The idea, as Douka explained, was to focus on the Mediterranean rim and try to trace migrations deep into Europe. There were two broad possibilities: one route was likely to have followed the Mediterranean coastline; the other would most likely have headed northwest, along the Danube River corridor. For many years people had assumed that the likely crossroads from Africa to Eurasia had been through the Near East but the evidence, and dating, for such a crossroads was scanty. Perhaps it was time that Palaeolithic archaeologists examined this again?

'For me,' Douka laughed, 'it was great timing. I had just started my PhD and I was interested in following up Mellars' ideas. So the idea was to date shell beads from around the Mediterranean to see if we could date modern human expansion. Traditionally Neanderthals are not thought to have used ornaments in the same way – or essentially not manufactured shell beads anyway. Ksar Akil was one of the first sites I looked at.'

I was keen to know more about the potential of such a simple and common resource as sea shells. 'Do you find these shell beads very frequently in modern human sites dating from this period?'

'Yes you do – especially around the Mediterranean. They are not rare findings, but they have exciting potential. And when you do find them, they are often in great numbers. We're talking about hundreds.'

'When you say around the Mediterranean – in what sort of places?'

'Lebanon, southern Turkey, southern Greece, Italy and all along the French and Spanish coasts extending also into the north of Spain.'

'What sorts of shells are we talking about?'

'Essentially they are small, about 1 to 2 centimetres diameter and perforated with a stone tool.'

'Too small to be considered food?'

'These are not food residue. These people were very selective in what species they gathered. Most likely they collected the shells directly from the beach as empty shells, but only freshly dead. They then pierced the shells – this is how we know that they were used as beads, whether as jewellery or decoration for clothing . . . So the idea for my research was to go back to the sites, go back to the original collections, and date this material, which amounted to several hundred shell beads.'

Carbon dating can make use of many different materials, including soft tissues, such as skin and derived leather, bone, charcoal, seeds – anything that exchanges carbon with its environment. Douka has made a special study of sea shells as a reliable source, which she applied to a new appraisal of the Ksar Akil site in Lebanon, working in association with colleagues, including Professors Robert E. M. Hedges and Thomas F. G. Higham from the Oxford Radiocarbon Accelerator Unit, Dr Christopher H. Bergman from Cincinnati, and Dr Frank P. Wesselingh, from Leiden in the Netherlands.

There were considerable problems that needed to be overcome, though. A major handicap was the fact that human fossil remains from the region had been scanty even in the original excavations, and much of this had been lost over the decades since. Excavation of Ksar Akil had begun as long ago as 1937 by a group of American Jesuits, and it had extended, in phases, and with different experts conducting the excavations, until 1975. The work over this period of time had revealed more than 30 stratigraphic layers some 23 metres deep. The earliest excavation had discovered the skeleton fossil playfully named 'Egbert'. The fossil is now lost and so it is only known through photographs and casts made at the time. When these casts were examined at the British Museum, by Bergman and Stringer, they suggested that the skull, which is small and delicately built, is likely to be that of a girl aged between seven and nine years. A second human fossil, that of a more primitive-looking upper jaw with a single canine tooth, was playfully named Ethelruda – perhaps after an Anglo-Saxon female saint. This was initially thought to be the jaw of a Neanderthal but is now considered very likely to be that of a modern human. While Egbert remains lost, Ethelruda was rediscovered in the Archaeological Museum of Lebanon but appeared to lack collagen, the usual source of carbon in such fossils which enables radiocarbon dating, so an alternative indicator was needed. Douka switched her attention to sea shells. As it happened, the molluscan collection from Ksar Akil was one of the largest ever discovered at a Palaeolithic site, containing approximately 2,000 specimens, the majority of which showed evidence of human modification such as perforation for ornament, or the presence of ochre pigments applied to them.

These shells had also been lost back in the 1960s, but in 2006 Douka and her colleagues traced them to the Netherlands. They were then able to locate the levels layer by layer, from the levels they knew were associated with Neanderthals to those associated with modern humans. The shells from these layers could now be transported to the Oxford laboratory for preparation and testing. Recent research into ways of excluding contamination when dating shell carbonates now proved useful in the preparation and grinding down of a small piece of each shell into a fine dust that she could use to radiocarbon date the site and, by inference, its associated human fossils.

'What dates did you find?'

'From the shell beads we got a radiocarbon dating for the earliest modern human layers at around 37,000 radiocarbon years, which converts to about 42,000 calendar years.' This allowed Douka to use a probability mathematical calculation, known as Bayesian modelling, to date the skull of Egbert to between 40,800 and 39,200 years old and the jaw of Ethelruda to between 42,400 and 41,700 years old.

This established time frame had additional implications; it put the arrival of modern humans into the Levant somewhat later than had been previously assumed. There was evidence elsewhere that the big migration out of Africa dated to several thousand years earlier than this. So Douka's findings suggested new possibilities. The main migration route might have been elsewhere, perhaps direct east from northeast Africa through Arabia and Central Asia, later curling towards the Near East. 'We needed to cast the net more widely to include other places and migration routes in the historic human dispersal out of Africa.'

Whether Ksar Akil had been on the crossroads, or whether the migration had arrived at Ksar Akil a few thousand years later, the site remains a very important source of information, bringing together hominin fossils with evidence of the prevailing culture. I had more questions to ask before moving further afield.

'What were these people like? What was their society like? What did they eat? What sort of clothes did they wear?'

'From Ksar Akil, and other sites along the Mediterranean, we know that these people were hunter-gatherers with a very varied diet. Throughout the year they were hunting large animals, such as deer, whatever was available. But they also ate a wide variety of different foods, including a wide array of plants, nuts and fruits. We are still working on this, but we think they may have been collecting shellfish when other dietary resources were sparse, probably in late winter or early spring.' In terms of population size, Douka thought we were probably looking at groups of about 80 people living in relatively close proximity.

But there was something else that puzzled me. While accepting that Ksar Akil was linked to the main movement of modern humans out of Africa and into Eurasia, I recalled that there was supposed to have been an earlier migration, judging from other evidence in the Near East. What, I wondered, had happened to that earlier migration? Why didn't they populate Eurasia? I knew, of course, that in Eurasia there had been a major problem with recurrent Ice Age pulsations, starting about 2,580,000 years ago and extending to the current interglacial period. These big freezes must have played havoc with human habitation and movements in Eurasia, in particular with a freeze called *Riss*, that lasted from 180,000 to 130,000 years ago, and one called the *Würm*, that lasted from 70,000 to roughly 10,000 years ago. Might this suggest that climate amelioration, beginning soon after the end of the *Riss* glaciation, had encouraged that earlier human migration out of Africa? But what then happened to that purported earlier migration?

*

Some of the best evidence for an earlier modern human migration out of Africa comes from the two fossil sites Es Skhul and Qafzeh. Es Skhul is a cave on the slopes of Mount Carmel and Qafzeh is a rock shelter in Lower Galilee. These two sites have proved rich in early hominin fossils, which have been tentatively dated at between 80,000 and 120,000 years old using techniques suitable for this level of antiquity, known as electron spin resonance and thermoluminescence. In particular the fossils include a number of well-preserved skulls of male and female individuals, and of a variety of ages. These skulls show a mix of archaic and modern features, with heavy brow ridges and a projecting lower face, similar to the Neanderthals, but the brain case and upper skull is more the shape associated with modern-day humans.

The mixture of features is so striking that at first these were interpreted as evidence for a partial evolution from Neanderthals to modern humans, but then obvious Neanderthal remains were discovered in the nearby Kebara cave that were dated to a much later period, roughly 61,000 to 48,000 years ago, disproving any such idea. Some anthropologists now proposed that the Es Skhul and Qafzeh hominins represent an early exodus of modern humans from Africa, around 125,000 years ago, so that the robust features represented a more archaic *H. sapiens* that would in time evolve to what we see today. If so, perhaps the Skhul/Qafzeh people may have represented a true earlier out-of-Africa migration that simply stopped in the Near East. In other words, a migration that came to a dead end here, with no contribution to the people of Eurasia.

'Yes,' Douka mused, 'we know there was an Ice Age coming. Perhaps in Eurasia at that earlier time, the Neanderthal population was at its peak. Perhaps they were holding strong. But then again it is very difficult to know why – it might just have been an effect of relative population sizes. Again, the earlier migration might not have become completely extinct . . .'

A set of teeth discovered in the nearby Tabun Cave, in 2005, was thought to show Neanderthal features. These teeth were tentatively dated to around 90,000 years ago. This, together with the discovery of more Neanderthal remains, again tentatively dated to 120,000 years ago, suggests that Neanderthals and modern humans might have made contact in the Near East at this time. This introduced the possibility that what we are seeing at Skhul/Qafzeh is a sexual crossing between the two species, giving rise to a hybrid people who were part modern human and part Neanderthal.

Dr Douka shrugged: 'This is something we are trying to look at. We know very little about any earlier migration from fossil evidence. I think a lot of weight is put into the Near East when it might not be the most likely place we should be looking at. But we now have a major five-year project under way that is looking at all the possibilities from Eastern Europe all the way to Central Asia, including Siberia.'

As we shall see in a subsequent chapter, some remarkable new genetic evidence is becoming available from the fossil remains of early humans, including Neanderthals, which is likely to revolutionise palaeoanthropology. It will clearly help when more archaic genomes from the fossil record can be added to the genetic picture, since these will not only allow genomic study for haplogroups but will also allow calibration of genomic dating with fossil-dating methods such as Carbon[14], electron spin resonance and thermoluminescence. In time, this will shed more light on the timing and migrations of our ancestors, whether they actually went north through the Levant or east through Arabia – or even directly across the Mediterranean Sea from North Africa. Some scientists are persuasively arguing that future study should include more of the extraordinary variation found in the human genome. The spread of endogenous retroviruses, such as HERV-106, HERV-113 and HERV-115, and the genetic and epigenetic regulatory regions of the genome, including those all-important non-coding RNAs, will also help to pin down movements and dates more accurately. I think we should be prepared for the true record to prove complex, with allowance for population advances and retreats, with mixing of different populations, and in particular with the problems of a harrowing Ice Age causing major upheavals in large swathes of the Eurasian landmass.

I asked Douka if we had any idea of the likely size of the migrating populations.

'I really don't think we have any true idea, because we don't really have much in the way of hard data that could lead us to calculate population size. But if you look at the genetic data coming out on the Neanderthal genome, it proposes an effective population size of between 1,500 and 3,000 reproductive Neanderthal women. Based on that you could be talking about a total Neanderthal population, in the European context, of roughly

10,000 individuals at the time the fossils were collected. If you ask me, I would hazard even less than that.'

'And contemporaneous modern humans – I presume they would also have run to small total populations?'

'Modern humans are considered a tropical species of humans, where population size is thought to have been much greater. Plus, the idea is that as they came out of Africa and dispersed across Asia, they maintained a wide networking of groups, with associated genetic exchange. So it is not unlikely that the population of modern humans had a higher turnover, and their numbers were larger.'

In a recent paper in the journal, *Science*, Mellars and his Cambridge-based colleague, Jennifer C. French, assessed relative populations of Neanderthals to modern humans using a combination of genetic and traditional archaeological techniques. They compared mitochondrial DNA diversity among present-day European populations to mitochondrial DNA diversity derived from Neanderthal remains. They also analysed various 'archaeological proxy evidence' for intensity of occupation over the Neanderthal to modern human transition in the well-studied southwest of France. They concluded that the numbers of modern humans settling the area showed a nine-fold increase on the numbers of Neanderthals that had previously occupied the same area. There are some obvious assumptions in such a technique, but their conclusions do pose an interesting question: what if the Neanderthal disappearance from Eurasia was brought about by simple numerical supremacy of arriving modern humans?

As we shall discover in the succeeding chapter, this, perhaps combined with some other emerging discoveries, might answer the question that has intrigued scientists and lay public alike for more than a century. But for the moment, we shall maintain our focus on the colonisation of Europe by arriving modern humans.

I pressed Dr Douka, 'So it seems that you get modern humans arriving into Europe – or Eurasia – and from that point on there appears to be a fairly rapid cultural evolution, perhaps starting around 20,000 years ago?'

'I wouldn't have put it at 20,000. For me it's probably earlier, more likely about 40,000 to 45,000 years ago. Some of these early modern humans would have occupied small pockets of Europe, southern Italy and western France, but after that, between 33,000 and 30,000 years ago, we really see something completely different. This marks the arrival of the Gravettians.'

'Another migration out of Africa?'

'We don't know where they come from or whether their culture first develops within Europe and then expands all the way to Russia, or vice versa. These humans appear around 33,000 years ago. And there is a completely new way of doing things. They bury their dead – and they bury them with thousands of beads. Some of these are shell beads but others are small ivory beads. They also use the canines of red deer, which they make into beads. They produced wonderful sculptural figurines. So we're talking about major cultural change.'

'We don't know whether it's an idea spreading, or whether it's a new people bringing in the new ideas?'

'That's right.'

'Perhaps in time the genetics will answer this?'

'Perhaps, yes. At present we have very little in the way of DNA from these people, but this is about to change. What we know, however, is that from 33,000 onwards there was a new blooming of human culture.'

These cultural and population changes took place at a time of monumental climate and ecological change in Eurasia. These severe climatic variations were accompanied by periods of major population movement and growth in the areas affected by the Ice

Age. A prolonged cold period, known in the scientific jargon as the Last Glacial Maximum, or 'LGM', occurred between 26,000 and 19,000 years ago. This cooling period is thought to have caused a massive population decline in Europe, with the survivors taking refuge in 'climate sanctuaries', or 'refugia', the four major examples of which were northern Iberia and southwest France, the Balkans, the Ukraine and the northern coast of the Black Sea and Italy.

The Last Glacial Maximum may well have reduced the genetic diversity of Europe, which in turn would complicate any assessment of arrival and dispersal based on the haplogroups of current Eurasian inhabitants. As the glaciers began to pull back, about 16,000 to 13,000 years ago, people began to repopulate the devastated landscape from the four geographic refuges, so that anthropological geneticists can expect to see not only genetic signatures that date from before the LGM but also the genetic signatures arising from the isolated populations of the four refuges, which would have expanded to fill the landscape as the glaciers melted.

For example, 80–90 per cent of males in Ireland, Wales, Scotland, the Basques in northern Spain and western French share the male-specific-region Y-chromosome haplogroup, or 'MSRY', known as 'R1b', as do 40–60 per cent of the male population of England, France, Germany and most of the rest of Western Europe. This makes it likely that their patrilineal ancestors took refuge in the northern Iberian climate sanctuary. In southeastern Europe R1b drops behind a related haplotype, R1a, in the area in and around Hungary and Serbia. Another MSRY haplogroup, labelled 'I', is found in its highest frequencies in Bosnia and Herzegovina, Serbia, Croatia, as well as Nordic countries, such as Sweden and Norway, and parts of Germany, Romania and Moldova. The same haplogroup clade is highly European-based, making some geneticists think it could date to before the last Ice Age. These, and many other different haplogroup clades, do help to trace populations and their movements, but I should add that, as we might expect from historic sources, they also show huge population mixing.

Curiously, when we look at the matrilineal-based mitochondrial haplogroups in Europe we see what appears to be a very different pattern from what we saw in the male-associated Y chromosome. When compared to the males, the female-associated haplogroups show much less geographic patterning, which seems to indicate that European women share a more common ancestry. How fascinating if this might reflect different socio-cultural traditions affecting the mobility of men and women.

Some 99 per cent of all European mitochondrial haplotypes fall within the categories H, I, J, K, M, T, U, V, and W or X. Haplotype H is the commonest, being found in no less than 50 per cent of all Europeans, and six of the above haplogroups, H, I , J, K, T and W are only found in European populations. The latter suggests that these haplogroups arose in the ancestral Caucasoid populations after they had separated from the ancestors of modern Africans and Asians. As historical records show, these ancestral genomes will have been blended with many different gene flows from eastern Asia, southern Siberia and Africa, but the ancestral patterns are still readily detectable even in modern genomes. Geneticists are currently assessing whether Europeans are mostly descended from Palaeolithic or Neolithic ancestors by looking at more and more genomes that date from 15,000 years ago or older.

These early ancestors were very similar to us, but they weren't the same. The use of genetic data to discern aspects of human prehistory is known as 'archaeogenetics'. One of the intriguing insights that is coming out of such archaeogenetic studies in recent years is the fact that significant evolutionary change has taken place in our human genome in the last 50,000 years – a key time in the human migrations out of Africa, and the colonisation by modern humans of Europe, Asia, Australasia and the Americas. Palaeoanthropologists have raised the possibility that this has been driven by the evolution of culture.

Perhaps we shouldn't be too surprised at this. Culture is a quintessential facet of human life and experience. Once again the geneticists have looked to mutational change, and in particular to the acquisition of new Snips that seem to mark out specific cultural populations. The theory is that some of these Snips in the auto-somal chromosomes are conserved because they happen to be close to a particular variant of a gene (in the jargon, a specific 'allele') that is already present, even in just a small number of individuals, perhaps dating back to a single ancestor, that has been favoured by natural selection because it gives a survival advantage.

Robert Moyzis and his colleagues at the University of California, at Irvine, searched for such evolutionary change among some 1.6 million Snips scattered throughout the entire human genome. They concluded that roughly 1,800 genes had been influenced in this way over the last 10,000 to 40,000 years – or, to put it another way, some 7 per cent of the genome had been specifically targeted by evolutionary forces during the expansion and settlement of humans in this phase of human history. The populations studied included Americans of European origins, Americans of African origins, and Americans of Asian (Han Chinese) origins. Key areas of the genome that appeared to be under intense selective pres-sure included some of the most important aspects of human internal chemistry. These included our ability to fight off infectious diseases, sexual reproduction, DNA chemistry and copying as part of the cell cycle, protein metabolism and the function of the nerve cells that form our brain and central nervous system.

The authors also concluded that we have undergone evolutionary change affecting many different physical attributes, very likely extending beyond those that they tested for. This study suggested that human evolution can involve widespread physiological and physical change in what, from an evolutionary perspective, would be a relatively short space of time. How likely is it that the same rapidly moving evolutionary selection pressures are still operating on us today?

*

As our journey progresses we become increasingly amazed at how the exploration of our human genome reveals so much that was formerly mysterious in our personal and cultural history. The potential for such exploration has undergone a sea change in recent years, presenting us with possibilities that would have seemed impossible a generation ago. As we shall now discover, it has become possible to learn from the study of genomes that diverged from our human ancestral line more than half a million years ago.

seventeen

Our Human Relatives

> *It's the questions and not the answers that are interesting,*
> *because these are questions that have no answers. But they*
> *are interesting questions to think about because they somehow*
> *reflect how we think about differences between us and our*
> *ancient ancestors.*
>
> SVANTE PÄÄBO

Svante Pääbo is a Swedish evolutionary geneticist who works at the Max Planck Institute for Evolutionary Anthropology in Leipzig, Germany. The son of biochemist Karl Sune Detlof Bergström, who shared the Nobel Prize in Physiology or Medicine in 1982 for discoveries related to the prostaglandins, Pääbo founded the new investigative scientific discipline known as palaeogenetics. Until recently geneticists had assumed that reading the genomes of extinct species of animals and plants, as popularised by Michael Crichton in his novel *Jurassic Park*, would be practically impossible because DNA degrades over time. The older the fossilised bones, the more degraded the DNA. Thus ancient fossils, dating back tens or even hundreds of thousands of years, were assumed to contain little or no residual DNA and were thus thought to be an unlikely source of useful genetic information. But over the decades since the 1980s, Pääbo and his colleagues at the Max Planck Institute began to make inroads into what had hitherto seemed impossible.

Through their work we have discovered that ancient DNA, though degraded, will sometimes survive the ravages of time. In pioneering this revolutionary new science of palaeogenetics, Pääbo has perfected methods of amplifying and then extracting genetic information from fossil bones and other ancient remains, making it possible to explore the genomes of long-dead animals and plants and, most intriguing of all, of ancient species of humans. In an interview with the online website, Edge, Pääbo admitted that he started out naïvely thinking it would be easy to study the genomes of long-dead individuals. To begin with his interest was the more recent past, and in particular the mummified bodies of ancient Egyptians, assuming these would be more amenable than the vastly older extinct hominins. He would subsequently confess: 'I was driven by delusions of grandeur.' But he persisted in his dream and, though it proved to be far from easy, he eventually succeeded.

Perhaps the simplest way to explain how he did so would be to climb back on board our train and pay a brief visit to the genomic landscape of such ancient fossils. Bone and tooth have turned out to be two of the best sources of ancient DNA in humans, so this will determine our choice of destination – the fossilised bone or tooth from a long-dead human. Here we enter a landscape very different from anything we visited before. We arrive at no neat railway line running away into the distance, east and west, but a confusion of fragments that, from a distance, resemble an explosion in a spaghetti factory. Pulling in closer, it still takes us a little while to recognise that what we are looking at is myriad haphazard shards of decayed genome. We are dismayed at the sight of these broken sections and pieces, which appear to be meaningless in terms of reading the original code in the pattern of sleepers. It seems to confirm what geneticists had long thought about the state of affairs in such ancient fossils . . .

But all is not quite as it seems. Each individual fragment of spaghetti contains a small but genuine fragment of DNA sequence – in our analogy a broken piece of railway track. And now, as we look at the vast scatter of fragments more closely, we see that each fragment contains anything from a few sleepers to perhaps a few hundred. Hardly reassuring, one might still be inclined to think, especially when the assembled human genome comprises 6.4 billion sleepers. Indeed, if all there were in this fossil bone was the fragmented remains of a single copy of the genome, the task of reading that genome would be hopeless. The key to understanding is that these myriad fragments are not bits and pieces of a single copy of the genome, these are the fragmented remains of vast numbers of the same genome, the residue of the billions of individual cells that made up the bone during life.

All of those different copies of the genome will have broken up into different fragments. And just as in the original Human Genome Project, when the genome was deliberately fragmented to reduce it to more manageable genetic bites, this huge variety of fragments, with breaks in different places, will contain DNA sequences that overlap. The question now is just how much of an overlap do we need to be sure that the overlap hasn't just resulted from chance? In fact, we can easily work out the mathematics with a simple calculation. What's the likelihood of, say, three, or four, or six, or eight nucleotides following an identical sequence in a row? Since there are just four nucleotides, the chance of the nucleotide matching gives us the random possibility of one in four – G, A, C or T. For the first two in sequence to be the same gives us a random possibility of 4 x 4 – a one in 16 chance. With each subsequent nucleotide we multiply by an additional 4. By the time we have calculated the mathematics for eight sleepers following identical sequences, the chance of this happening by accident is one in 65,536 – in other words, extremely unlikely. We now have a system that works.

The first step is obvious. We need to sequence every fragment and bank this information in a computer with a very large memory. The second step is to search the banked sequences for matching sections that will pick out areas of overlap between the different fragments. From there we can begin the laborious process of stitching sequences together using the overlaps to identify contiguous areas. In fact, we are reproducing pretty much the way the first draft of the human genomic sequence was calibrated, except we are dealing with a vast collection of much smaller sequences that must now be painstakingly knitted together.

To this breakthrough Pääbo enlisted two others. The first of these was Kary Mullis's Nobel Prize-winning discovery of the polymerase chain reaction. This ensured that even if the individual fragments were present in very few copies, too tiny by far to be detectable by the automated sequencing machines, they would be duplicated over and over until they became identifiable. Next came the brainwave of teaming up with an innovative American biotechnology company, 454 Life Sciences, which had been set up by Jonathan Rothberg to develop machines that were capable of automated high-throughput DNA sequencing. The company had been taken over by the pharmaceutical giant Roche in 1997. The entire process could now be automated. The machines could automate the extraction of ancient DNA, amplify it using PCR, and thus generate a soup of small bits and pieces of the entire genome, from which they could bank and then knit it all together by aligning the overlaps.

Of course, there were other problems to be overcome, such as contamination of the test samples by unwanted sources – human, animal and bacterial. They found ways of removing contaminating DNA with enzymes. They also avoided contaminating specimens with their own DNA by working in a designated 'clean room' and by taking strict biosafety precautions. During the actual sequencing they discovered that one of the DNA nucleotides, cytosine, sometimes degraded to uracil, the nucleotide that replaces thymine in RNA. Then it became even more complicated; sometimes cytosine that had been epigenetically tagged with a methyl group degraded not to uracil but to thymine. This created confusion until they realised that, by accident, they had made a significant discovery. In 2009 one of the team, Adrian Briggs, figured out a method for distinguishing thymines derived from methylated cytosines from original thymines. Now they could read off some of the original epigenetic programming of the genome. So it happened that, little by little, the scientists advanced and perfected their DNA extraction techniques and improved the quality of the readouts.

The applications were taken up by other laboratories and extended to the woolly mammoth, cave bear and coelacanth, as well as fossil plants. Meanwhile Pääbo and his team addressed the single most exciting challenge that had faced his lab from the very beginning of this pioneering research: they would attempt the extraction of the genome of our long-lost human cousin, Neanderthal Man. Time, perhaps, for us to take a brief look at what is known of our human evolutionary history.

*

In his notebook, during the decades in which he gradually assembled his thoughts on how a new species might arise from an existing one, Charles Darwin sketched something resembling the branching pattern of a tree. This branching pattern has since then been amply confirmed by biological study and the same pattern neatly fits with the Linnaean classification of life into species, genera, families, and so on – the disciplines biologists call 'phylo-genetics' and 'taxonomy'. Humans are no exception to this; as recently as 70,000 years ago some five different species of humans cohabited the Earth. All five had directly or indirectly descended from a single ancestral species, known as *Homo erectus*. One of the five was our direct ancestors, 'early modern humans' – or to give them their formal Linnaean classification, *Homo sapiens*. Both *H. sapiens* and the still-evolving *H. erectus* shared the planet with Neanderthal Man, or *Homo neanderthalensis*, who inhabited large areas of Eurasia; the so-called 'Hobbit', or *Homo floresiensis*, who lived on the Asian island of Flores; and a mysterious species only recently mooted, called Denisovan Man, or *Denisova hominins*, who appears to have inhabited parts of Asia. All members of the human evolutionary tree that evolved after our divergence from the common human-chimpanzee ancestor are referred to as 'homi-nins'. We need to distinguish this term from a second umbrella term, 'hominids', which includes all hominins as well as all modern and extinct Great Apes, including chimpanzees, gorillas and orang-utans.

A good place to begin our exploration of this human evolu-tionary tree is with the common ancestor *Homo erectus*, who first evolved in Africa approximately 2 million years ago. An almost complete skull of an early individual was unearthed in Kenya and an almost complete skeleton of a juvenile was discovered near to Lake Turkana, Kenya, by Richard Leakey's team. This juvenile skeleton was attributed to a boy who, judging by his five-foot-three stature and skeletal development, was initially thought to be 12 or 13 years of age. Later on his age had to be adjusted to eight years when scientists discovered the meticulous accuracy of counting the growth lines in teeth. This intrigued scientists since it suggested that childhood in *Homo erectus* was much shorter than that of human children today.

Homo erectus individuals were as tall as modern humans and they were even more robustly built. They controlled fire in hearths and manufactured stone tools of the 'Acheulean' set, including beautifully sculptured hand axes and cleavers, which demanded the ability for mental imagery and planning and would have enabled them to kill and butcher animals. There is also some evidence that they may have cared for their weak and elderly. The skull of *H. erectus* was primitive in its features, with a heavy brow ridge, a low, flat vault and protruding jaws, including a poorly developed chin. The brain volume was about twice the size of a chimpanzee's, at approximately 850ml, which compares to 1,300ml in an average male today. *Homo erectus* continued to evolve within Africa but, rather surprisingly, also migrated out of Africa at a very early stage in its evolution to populate the Eurasian landmass in the earliest-known hominin migration. Famous *H. erectus* fossils include skulls and jaws from Dmanisi in Georgia, dating to roughly 1.8 million years ago. Other fossil evidence finds *H. erectus* colonising Asia close to a million years ago and the Middle East and Southern Europe about 730,000 years ago, where, as in Africa, the species continued to evolve bigger brains long after it had spawned the four descendant species, including our own. From this original dispersal there are two schools of thought as to how modern humans might have come on the scene.

The 'multi-regional theory' pioneered by Milford H. Wolpoff suggests that subsequent human evolution of *Homo erectus* in regions such as Africa, Europe, Asia and Australia gave rise to distinct local lineages, albeit with some mixing, through mating, between regional groups and perhaps with peoples migrating out of Africa. The 'out of Africa' theory proposes that all currently living humans are descended from early modern human ancestors that evolved in Africa roughly 180,000 years ago, and who migrated, perhaps in a series of waves, out into the rest of the world. Until recently these competing evolutionary theories were largely based on the fossil record and their associated palaeoarchaeology. As we have seen, the haplogroup genetic tracing tends by and large to favour the out-of-Africa scenario, but this leaves open the possibility of subsequent admixture of modern humans with other regionally distributed species of humans that had evolved from earlier migrations out of Africa, thus perhaps adding in a regional aspect to our evolution.

Palaeontologists had assumed that it was exceedingly unlikely that they would ever have the chance to study the genomes of these ancient cousin species, but now the picture was radically changed by Pääbo's spectacular breakthrough in genetic technology, with its pioneering of a whole new arm of genetic investigation – including a new name for it: palaeogenetics.

In interviews, Pääbo admits that his work in relation to palaeontology was inspired by the desire to answer many questions. To begin with, was it possible to extract any DNA from such ancient fossils? Even if he managed to extract useful sequences, perhaps confined to mitochondrial DNA, what would this tell us about our human evolution? What if he could extract significant nuclear genomic sequences? Would this help to clarify the debate as to why the Neanderthals became extinct? Would it also answer important unknowns about our own species' evolution? For example, would it help to clarify which of the competing hypotheses for our own human origins was correct?

Many of the most interesting and provocative questions were directed at the Neanderthals. What do we really know about them? Were they the stupid cavemen of popular prejudice? In what landscape and ecology did they live? How did they survive in terms of food and shelter? What was their society like? Would they have understood love, family, friendship? How were they like us – and how did they differ? What possible catastrophe could have so befallen a people who had survived for a quarter of a million years in the Eurasian landscape only to become extinct within ten to fifteen thousand years of the arrival into Eurasia of modern humans?

*

Neanderthals are so named because one of the early fossils was found in a cave in the valley of the Neander River, in Germany. The 'th' diphthong is pronounced 't', in the German manner, so Neanderthal is pronounced 'Neandertal'. The earliest features typical of Neanderthals first appear in the European fossil record about 400,000 years ago, which probably marks something close to their evolutionary origins. They are generally thought to have descended from an intermediary species, *Homo heidelbergensis*, which in turn descended from *Homo erectus*. Neanderthal fossils and tools are found widely throughout Europe and western Asia, as far east as southern Siberia and as far south as the Middle East, before disappearing from the fossil record about 28,000 years ago, with the most recent remains found in a cave in Gibraltar. And although they are usually classed as a separate species from modern humans, evolutionary biologists regard them as our closest and most recent human cousins.

What did they look like and how do they compare to us?

On average they appear to have been shorter and stockier than us, in particular having shorter legs and forearms. Their stature is thought to be an adaptation to living in a cold climate. Their skulls were long, more flattened on top than our own. They had much more prominent eyebrow ridges and their noses were also much larger than we find in the average modern human – Stringer imagined that their noses must have been 'remarkably prominent'. In some cases the nasal bone jutted out nearly horizontally under the brows. Their front teeth were very large and often heavily worn even when compared with those of their own ancestors, *H. heidelbergensis*. Stringer and Gamble have queried if these facial appearances are an adaptation that came about because the Neanderthals may have used their incisor teeth as an extra appendage. Many adult Neanderthals display incisors worn down to mere stubs, suggesting they might have used them like a portable vice while working some material, perhaps, in the opinion of Shara Bailey at New York University, the processing of skins to make leather goods. Their cheeks were swept back on either side of the nose, giving the central part of the face a marked protuberance. The lower jaw followed this forward projection of the upper face and, like *H. erectus*, this resulted in the loss of a pointed chin. When compared to modern humans, Neanderthals also showed differences in their chest shape, the pelvic bones and the limb bones, which tended to be thicker, with wider, stronger joints, all of which suggest that the Neanderthals were adapted to more powerful physical activities and stresses than was the case with early modern humans.

Contrary to the unenlightened earlier interpretations, Neanderthals were neither stupid nor brutal. The Neanderthal brain is slightly bigger in volume than our modern human brain, even today. Casts of their brains taken from fossil skulls show that they had the same tendency to be right-handed. Intriguingly, their eye sockets were larger than ours and the occipital lobes of their brains were also larger, suggesting that, perhaps, they had better night vision than us, which might have given them a survival advantage in their hunting lifestyle in the murky twilight of the cold northern climate, with its dark and dreary winters. Another intriguing difference was the time taken for childhood growth and maturation. As we saw with *H. erectus*, studies based on dental growth lines suggest that Neanderthal children may also have matured more quickly than modern human children.

We need to avoid drawing derogatory inferences from such observations. For example, we should compare Neanderthal children's growth not to that of modern human children but to the fossils of early modern children, dating to the same time period as the Neanderthals under study. But if this proves to carry true, this shortened childhood might have important cultural implications since so much learning takes place during the extended period of development of our children.

One of the more questionable theories to emerge in the past has been the suggestion that Neanderthals did not possess language. They must have had language, albeit it would very likely have been simpler than we humans possess today. They lived in hunter-gatherer bands of perhaps a dozen or two, very likely extended families. Although few Neanderthals appear to have lived beyond 40 years, there is some evidence that they possessed knowledge of herbal medicine, cared for their elderly and looked after infirm individuals.

We are currently in the middle of an extensive reappraisal of Neanderthal evolution and culture in which some palaeontologists draw attention to the fact that the Neanderthals survived in Europe for roughly 250,000 years, despite very testing climatic conditions, colonising a vast geographic territory. And now, as the first breakthroughs in genetics became available, we find that they were not swarthy and covered with black hair, as in the early drawings and models, but more likely fair-skinned, as we might have better imagined a species long adapted for survival in the cold northern climates of Europe. Still, the die-hards insisted that Neanderthals lacked the one key feature that lay at the very core of our advanced civilisation: they lacked the higher cognitive function that enabled complex language and symbolic thinking, the quintessentially human cognitive breakthroughs that enabled symbolic art and subtleties of reasoning that came with language.

In 2001, a human gene was identified by Oxford University geneticists Simon Fisher and Anthony Monaco, now known as FOXP2, which is important in our ability to articulate language. Mutations of the gene gave rise to difficulty with muscle control in the vocal cords, tongue, and lips that were needed to speak. This made scientists wonder if this particular gene was a key acquisition exclusive to modern humans and responsible for our evolution of language.

However, a single gene is unlikely to code for our ability to speak. The evolution of speech and language involved complex changes in the structures of the human voice box, or larynx, the throat and the mobility of the human tongue and lips, all of which would also have required long-term and complex modification of the areas of the brain that control thought, sensation, movement and the coordination of these various body parts. We know that speech is dependent on the development of specific regions of the brain, such as Broca's area, which would hardly have evolved over the 200,000 years of our separation from Neanderthals. From study of feral children, we know that so closely is this 'speech module' linked to culture that if a child is not exposed to picking up language from parents and those teaching him or her up to around the age of seven, they never develop proper speech. Remarkably, the late neurobiologist and avian scientist, Peter Marler, discovered that something similar applies to the song of birds.

Socially there is some evidence that the Neanderthal hunter-gatherer bands were smaller in number than those of early modern humans, and they appear to have been less mobile and possibly less connected with other population groups. Their tools were adapted from local resources where the early moderns appeared to trade, and move, more widely. At first anthropologists believed they did not produce original art, in terms of cave painting or personal ornaments, such as necklaces and sculptured items. But a modern reappraisal suggests that the Neanderthals were more inventive than previously assumed. They made use of string some 90,000 years ago. The presence of Neanderthal tools on Mediterranean islands confirmed that they knew how to cross the sea in some kind of craft. Where the appearance of complex tools and shell or bead body adornment in Neanderthal settlements were thought to have been copied from arriving early moderns, a collection of sophisticated bone tools, known as lissoirs, were discovered at two sites in southwest France that were dated to between 45,000 and 51,000 years ago, which, if true, is several thousand years before early moderns arrived there. Some anthropologists still proposed that, on the weight of evidence, Neanderthals lacked the higher mental evolution implicit in symbolism, pointing to the beautifully illustrated cave paintings of France, Spain and much further afield in Australia; but others have cautioned that the wonderful cave paintings were not produced by early modern humans at the time of the Neanderthals, but painted some 20–30,000 years after the Neanderthals had disappeared – plenty of time for a subsequent cultural evolution.

In 2010, a group of scientists from many different university centres published a paper on the symbolic use of marine shells and mineral pigments by Iberian Neanderthals. Their paper, headed by João Zilhão, reported the discovery of two Neanderthal-associated Middle Palaeolithic sites, one a large cave near to the Mediterranean and another a rock shelter by the Mula River, both in present-day Spain and dated to roughly 50,000 years ago. Here the investigators discovered pigment-stained marine shells that had been deliberately perforated to allow them to be strung together into some kind of decorated adornment. They also discovered lumps of yellow and red colorants and residues preserved inside a *Spondylus* shell, together with many other tools and artefacts. The symbolic use of perforated and pigmented shells, used for necklaces, contradicted previous cultural assumptions. It suggested that Neanderthals may have had a similar ability for symbolic thought to early modern humans – they might even have taught our arriving early moderns a thing or two about the sophisticated manufacture and use of bone tools.

As part of this reappraisal some palaeoanthropologists are coming round to a new perspective on the Neanderthals, arguing that they were little different in their cultural evolution from early moderns at the time they first emerged from Africa. For example, two American experts, P. Villa and W. Roebroeks, suggest that much of the earlier prejudice had resulted from a false comparison between Neanderthals, based on dig sites dating from the early to middle Stone Age, and modern humans from much more recent times. Others argue that even if modern humans were more advanced in their culture than Neanderthals, this does not necessarily reflect superiority in genetic or intellectual potential, but the sort of differences in culture seen in our modern human history, brought about by the development and transmission of ideas.

However, none of this debate reveals why the Neanderthals vanished off the global map, facing us with the mystery of their disappearance. Were they exterminated in the same way that some native populations were exterminated by arriving European colonists? Were they wiped out by diseases carried into Eurasia by moderns arriving out of tropical Africa? Were they simply outcompeted through the cultural superiority of the arriving modern humans? These questions had dogged palaeoanthropologists for a century and a half. Svante Pääbo's new discipline of palaeogenetics was about to suggest a completely different explanation for the extinction of the Neanderthals.

eighteen

The Fate of the Neanderthals

> *Early in our analysis of the human remains from Lagar Velho, we proposed that the child's skeleton presented evidence of prior blending of local Neanderthal and arriving early modern human populations in western Iberia. Our interpretation has been widely accepted as both interesting and reasonable, being rejected* a priori *only by those who are intellectually immune to the idea of Neanderthal–modern human productive interbreeding.*
>
> JOÃO ZILHÃO AND ERIK TRINKAUS

There are, as we have seen, two quite different genomes within each human cell, the mitochondrial and the nuclear. Since there are hundreds of mitochondria in every cell, each with its own bacterium-derived genome, by far the most copious DNA in a fossil bone is going to be mitochondrial. The mitochondrial genome is also much smaller than the nuclear genome. So the logical place to begin the exploration of the Neanderthal genome was with the extraction and decipherment of the mitochondrial DNA.

After some preliminary sequence discoveries over a number of years, in 2008 Pääbo and his team at the Max Planck Institute, working in cooperation with colleagues from America, Croatia and Finland, published the first complete sequencing of Neanderthal mitochondrial DNA, which they had extracted from a 38,000-year-old fossil bone. The discovery was groundbreaking, not merely in the historic sense but also the significance of their findings: 'It establishes that the Neanderthal mitochondrial DNA falls outside the variation of extant human mitochondrial DNAs.'

This confirmed the long-held view that Neanderthals were a different evolutionary line from modern humans. In applying the customary search for nucleotide polymorphisms, Green and colleagues discovered far more Snip-type differences between Neanderthals and humans than we see across the global modern human divide. Given that mutations occur at roughly predictable intervals, as in the mutational clock, '[It] allows an estimate of the divergence date between the two lineages of 660,000 plus or minus 140,000 years.' This came close to the divergence date previously proposed by palaeontologists on the basis of fossils and archaeological findings.

The paper also predicted something that might prove helpful in working out the fate of the Neanderthals after the arrival of early modern humans into Europe, the Middle East and Asia: 'There is evidence that ... the effective population size of Neanderthals was small.' By effective population size, the scientists were referring to the Neanderthal species gene pool – or, to put it another way, the genetic diversity of the Neanderthal species. We need to be cautious in interpreting something as sweeping as this from a first draft of the relatively small, and entirely female-associated, mitochondrial genome. But if this were to be confirmed from study of the nuclear genome, it would have important implications for the impact of interbreeding between Neanderthals and modern humans. For the moment let us flag this up as something to keep in mind. Meanwhile we should consider the suggested timeline for the separation of the modern human and Neanderthal lineages.

Both the palaeontological and genetic evidence is pointing to a divergence of the modern human and Neanderthal lineages roughly half a million years ago. Based on the study of other mammals, this is a relatively short time for complete separation into two reproductively distinct and separate species, which is generally thought to require something like a million years. So let us flag that up as a second consideration to keep in mind. Some readers might have realised the oddity posed by this timing of divergence; if the Neanderthals and modern humans began their separate evolutionary trajectories roughly 500,000 years ago, how do we explain the purported African origins of the modern human evolutionary line at 180,000, perhaps at most 200,000 years ago?

In fact there is no contradiction. There is a common consensus that both species evolved, albeit through different lineages, from a common ancestor, *H. erectus*. For the first 300,000 years after separation, both lineages would have continued to evolve, presumably without contact with one another, modern humans in Africa and Neanderthals in Eurasia. And thus, by 180,000 years ago they had become sufficiently different to be considered distinct evolutionary lineages, heading towards becoming separate species. In Neanderthal terms, there is a general consensus that the intervening interval between *H. erectus* and the Neanderthals was filled by the intermediate ancestral lineage of *H. heidelbergensis*. This also leaves open the question as to whether *H. sapiens* evolved from *H. heidelbergensis* – albeit given a different name – in Africa.

Puzzled about this, I wrote to Professor Stringer, the human origins expert at the Natural History Museum in London, and essentially he confirmed that there were indeed two alternative theories. One theory proposed that both the Neanderthal and the modern human lineages went through the intermediate step of *H. heidelbergensis*, probably separately in Africa and in Europe. The other theory proposed that only the Neanderthal lineage went through the *H. heidelbergensis* stage in Europe: meanwhile *H. sapiens* evolved directly from *H. erectus* in Africa. Here then is another situation that we might keep in mind with respect to any future genomic revelations.

All along Pääbo had made clear that it was his ambition to reconstruct the entire Neanderthal nuclear genome. In 2007, a year before the mitochondrial announcement, and the same year he was named one of *Time* Magazine's 100 most influential people of the year, European researchers reported that Neanderthals had a gene mutation that was associated with the fair skin seen in modern Europeans. This same mutation may have given some of them red hair. The following year, another paper reported that Neanderthals had the same O blood group as modern humans and they shared the language-associated gene, *FOXP2*, with us. In 2009, at the annual meeting of the American Association for the Advancement of Science, it was announced that the Max Planck Institute for Evolutionary Anthropology had completed the first rough draft of the Neanderthal nuclear genome, which included some four billion of the estimated total of 6.4 billion nucleotide 'sleepers' of DNA extracted from the fossil bones of three individuals. This was duly published in 2010.

The first and the predictable outcome was that more than 99 per cent of the Neanderthal genome was identical to our own. This was hardly surprising since we had evolved from a common ancestor only half a million years earlier. Much of our genome, and in particular the protein-coding portion, has to do with everyday aspects of our inner chemistry, cell structures, cell repair, cell death and removal, and cell regeneration, as well as immunological determination of self and the constant battle against microbial disease. We would expect these portions of the two genomes to be largely identical. A more surprising discovery was that the Neanderthal genome showed more commonality with modern humans who had evolved in Europe and Asia than with modern humans who had evolved in Africa. In scientific parlance, 'a parsimonious explanation for these observations is that Neanderthals exchanged genes with the ancestors of non-Africans'. In ordinary English, our modern human ancestors interbred to a significant degree with Neanderthals. To put it more bluntly still, some of our distant grandparents were Neanderthals.

The findings suggested that people of European origins inherited between 1 and 4 per cent of their nuclear genomic DNA from Neanderthal ancestors. The preliminary evidence suggested that people of Asian origins may have inherited even higher percentages of their DNA from their distant Neanderthal ancestors. As the paper expressed it, Neanderthals appeared to be as closely related to Chinese and Papua New Guineans as they were to Europeans, despite the fact that Neanderthal fossils have only been found in Europe and western Asia. Perhaps early moderns had already mixed with Neanderthals part way through their migration, taking their Neanderthal ancestry with them in their expansion into Eurasia.

Just how commonplace was the interbreeding between the two populations?

Previous studies of hybridisation in evolutionary biology had suggested that when a colonising new population encountered a resident population, even a small number of breeding events along the wave of interaction can result in substantial introduction of new genes into the colonising population. It's all a question of population numbers and subsequent population expansion. It seems that the incoming genes – Neanderthal in this situation – can 'surf' to high frequency as the colonising population expands. There is, of course, an alternative, simpler explanation. Perhaps interbreeding between modern humans and Neanderthals was truly commonplace?

The news that Europeans and Asians were part Neanderthal in their genetic origins hit the popular media like a bombshell. What should have been anticipated, really, given human behaviour, now provoked astonishment.

*

From the first discovery of their fossils, Neanderthals had been the subject of fascination as a rival species to our own – but they had also been the subject of enormous prejudice, and this, alas, was true of professionals and lay media alike. Marcellin Boule, the first expert to review their fossil remains, painted a picture of a brutally apish creature in which there could only be 'rudimentary intellectual faculties' and in which 'all traces of any preoccupations of an aesthetic or of a moral kind' were unimaginable. This pejorative pattern dominated the opinion of Neanderthals for the first half of the twentieth century. It was an unfortunate accident of fate that Boule's post-mortem examination featured the partial skeleton of an older male who had been badly maimed in some accident, provoking severe secondary osteoarthritis. This prejudicial interpretation from such a distinguished source influenced opinion far beyond the objective world of archaeology and palaeontology.

H. G. Wells wrote a story featuring the deserved extermination of what he labelled the 'grisly folk'. While more recent writers, such as William Golding and Jared Diamond, presented a more sympathetic view of the Neanderthals, they also assumed their extinction at the hands of our more advanced ancestors, even though there is no hard palaeontological evidence to support it.

This bigotry prevailed until relatively recently, with anthropologists arguing that Neanderthals couldn't speak, or if they did so it was in a peculiar voice, this in spite of the fact they had brains slightly larger than our own and the speech area of their brains, known as Broca's area, was little different too. The large brain was downplayed as poor in quality. Their tools, typical of what is called the Mousterian culture, were less sophisticated than those of modern humans. In particular they appeared to lack throwing spears, compelling them to face up directly to the huge prey they hunted, such as woolly mammoths and rhinoceroses.

There were, of course, very real differences in skeletal morphology between Neanderthals and early modern humans, but researchers chose to focus on the differences while ignoring the similarities. In the last two decades a growing number of palaeontologists have re-evaluated the evidence for Neanderthal society and culture, concluding that preconceived notions may have led earlier researchers to ignore the actual evidence for Neanderthal inventiveness and culture. Their brains had large frontal lobes – the regions associated with intelligence and culture. Ralph Holloway at Columbia University in New York has studied hundreds of brain casts from Neanderthal skulls to confirm that they had the same level of development of the speech area as we do. Sites in France have confirmed that Neanderthals didn't just occupy caves and rock overhangs but they erected shelters, leaving traces of the supportive wooden posts. Their tools are now seen as difficult to manufacture, involving planning, vision and great skill. There is growing evidence that they wore clothes and that they adorned themselves with symbolic artefacts, including perforated and painted shells. They buried their dead, perhaps with ceremony, and may have played music. And although they hunted formidable prey, such as rhinos, mammoths and bison, they also adapted their strategies to suit different environments, trapping and hunting birds and rabbits and gathering seafood.

Penny Spikins, and her colleagues at York University, identified at least three sites: Wansunt Road in Kent, Foxhall Road in Ipswich, and Rhenen in the Netherlands, where miniature Neanderthal hand axes, which were likely to be children's toys, have come to light. In Arcy-sur-Cure in France, and a site in Belgium, other palaeoarchaeologists have discovered collections of expertly chipped stone tools next to some very inexpert attempts, which suggest Neanderthal adults teaching children how to make them in a Stone Age equivalent of schooling. The role of emotions, such as compassion, is central to the behaviours and intimate relationships that define human society, but palaeontologists have been cautious about extrapolating emotional significance to archaeological findings. Spikins and her colleagues have attempted to explore scientific constructs that might offer a basis for such studies, looking for the evidence of empathy and compassion in archaeological contexts from the earliest archaic humans to modern society. They found such compassion, in caring for sick and injured individuals in Neanderthal sites from 'the Old Man of Shanidar', who suffered from terrible injuries, to the Sima de los Huesos – the pit of bones – where a child suffering from a hereditary disease affecting the skull was cared for up to its death at the age of eight.

Given this accumulation of evidence, Erik Trinkaus at Washington University in St Louis, Missouri, has concluded: 'If you look at the archaeological evidence of Neanderthals and modern humans . . . they are very similar. Neanderthals were people, and they probably had the same range of mental abilities we do.'

Of course, not everybody agrees. Some distinguished researchers, such as Mellars, think that Neanderthals, while skilled enough to survive in their Eurasian landscape for more than 200,000 years, were less cognitively equipped than contemporaneous modern humans. He argues that by the time modern humans arrived into

Europe, they had better technology, better social organisation and, by inference, better brains. Steven Mithen, at the University of Reading, agrees with him. They may be right. However, we need to distinguish social evolution from the hereditary ability for intellectual thought. We also need to be careful to compare like with like – in other words, to compare Neanderthal culture with modern human culture during the same era. One only needs to reflect on the massive cultural differences between different modern human populations a century or two ago – which were more associated with education and the inheritance of ideas than differences in heredity or inherent intellectual ability between populations.

The encouraging momentum that is now sweeping like a tsunami of potential adjudication through such passionate debate is that, thanks to Pääbo and his discoveries, the hard facts of genetics can now be combined with the equally hard facts of archaeological dating techniques, and it will be facts and not prejudicial supposition that will ultimately define future belief.

Meanwhile, there are some relevant hard facts that should be taken fully into account in the fast-developing scenario.

*

Hybridisation, or sexual crossing between dissimilar species – or subspecies – is one of the four major mechanisms that give rise to the hereditary genetic change that makes evolution possible. It can, where there are major genetic differences between the hybrid partners, give rise to serious genetic dysfunction, including infertility. Scientists who study hybridisation in animals and plants have found that the closer the evolutionary lineages involved, the shorter the time of separation from a common ancestor, and thus the less different the two genomes and therefore the more stable the genetic outcomes. The genetic interbreeding between Neanderthals and our modern human ancestors hardly gave rise to infertility since we are the hybrid descendants. And the Neanderthal component of our Eurasian genomes today is not small. At its upper range of 4 per cent, this is the genetic legacy you or I would have inherited from a great-great-great grandparent from just a century ago. At DNA level, this is huge. One need only divide 6.4 billion sleepers by 25 to see how many sleepers our train would have to traverse in the exploration of this genomic contribution – 260 million – all making up distinctive genes, viral sections, regulatory sequences and non-coding RNAs.

Some authors appear to be missing the point in interpreting such a major 'hybridisation event'. The pooling of two different evolutionary lineages will give rise to an immediate dramatic increase in genetic diversity in the hybrid offspring, an increase that will be inherited by the descendants in all future generations.

We might consider what this implies: that the two genomes contain differing genetic histories, including genetic adaptations, with potential advantages for survival that have been hard won in different ecologies and environments. Studies of the effects of such 'hybrid creativity' in nature suggest that the hybrid offspring may be better able to withstand tough environmental conditions, including harsh climates, than the non-hybrid parental lineages. Natural selection will not choose one lineage over another. Since both lineages are now common to a single genome, selection will now work at the level of the hybrid genome, just as it worked at the level of the holobiontic genome in cases of genetic symbiosis. Genetic sequences that impair survival, regardless of their species origins, will be selected against; meanwhile, genetic sequences that enhance survival, regardless of origin, will be favoured. The fact that we still retain a huge Neanderthal component to our genomic make-up speaks for itself. It strongly suggests that our ancestors did gain an advantage for survival in the sexual crossing with Neanderthals. Moreover, it isn't hard to see why this might be so.

Our modern human ancestors had evolved in the tropical heat and bright sunshine of Africa. They encountered Neanderthals when moving into a much colder, less sunny environment in which the winters in particular would be long, cloudy and gloomy. One key effect of this would have been a failure to make enough vitamin D in their darker skins. Given the effects of vitamin D deficiency in softening bones, or in weakening the immune system, this Neanderthal contribution would have given the hybrid offspring a better chance of survival.

But the benefits of cooperation between the two populations may well have extended far beyond increased genetic diversity. How likely was it that there would have been social and cultural advantages of exchange between the two different populations? This was no meeting of Europeans bearing the fruits of Renaissance enlightenment, or the Industrial Revolution, into a Stone Age hunter-gatherer society. Both populations were still at the Stone Age hunter-gatherer stage. The exchange of knowledge of the local geography, of flora and fauna, of seasonal availability of foods, sources of shelter, perhaps evolved techniques of working skins or other items of clothing, herb lore – and, perhaps, more symbolic aspects of culture, such as decoration, music – would have worked both ways.

*

During the search for Neanderthal fossils that grew up around Pääbo's genomic exploration, Russian archaeologists had been digging in the floor of a gigantic cavern high up in the Altai Mountains of southern Siberia. The climate on the northern slopes where the cave is situated is arid and cold, with a mean year-round temperature of about zero degrees Centigrade. In 2010 the archaeologists discovered a toe bone from a Neanderthal female from a layer dated to approximately 50,000 years ago. This bone proved to be incredibly rich in archaic DNA, enabling Pääbo's group to extract the best-quality Neanderthal nuclear genomic sequences to date. In January 2014 scientists from many different centres and laboratories combined forces with Pääbo's group to analyse and report the complete Neanderthal nuclear genome of this individual, comparing and contrasting it with the human genome and with what was already known from the previous examination of Neanderthals from widely different geographic localities, including three Neanderthal individuals from the Vindija Cave in Croatia and a Neanderthal infant from the Mezmaiskaya Cave in the Caucasus.

Were we to return to our magical train, would we recognise the sections of track that we owe to the Neanderthal part of our heritage?

In March 2014 three different Harvard-based groups of evolutionary geneticists, including David Reich and colleagues based at the Department of Genetics at the Harvard Medical School, and groups at the Broad Institute and Howard Hughes Medical Institute, combined forces with Pääbo's group at the Max Planck Institute in Germany to publish an overview on the genomic contribution of Neanderthals to present-day humans. In the opinion of these authors, the answer to my question is a definite yes. Although a lot of time has gone by since the hybridisation event, the Neanderthal chunks of track are readily distinguishable because of their haplotypes. Where at the time of hybridisation fully half the genome of the first offspring would have been Neanderthal, the dilutional effects of time and a succession of within-species matings has reduced the Neanderthal sections to sequences of DNA less than 100,000 sleepers long – small perhaps against the overall size of the genome at 6.4 billion sleepers, but enough to offer a series of lengthy journeys to our magical steam train. In the words of the authors: 'Neanderthal haplotypes are distinctive enough that several studies have been able to detect Neanderthal ancestry at specific loci.' What then did they actually discover when they visited the genomes of some 1,004 present-day humans, examining the track in minute detail and taking a long hard look at these regions that are identifiably Neanderthal?

The Reich group concluded that regions of our modern genome that are particularly rich in Neanderthal genes included those coding for the proteins that make keratin filaments. Keratin is the structural material that makes up the outer layer of our skin, and in modified form is the main structural component of our hair and nails. One of the specific genes they may have bequeathed us is *BNC2*, which is involved in skin pigmentation. Researchers based in the University of Arizona also discovered that a proportion of Eurasians, and especially Melanesians, had inherited the genetic region known as *STAT2* from Neanderthal ancestors. *STAT2* is part of the system that determines the identity of self and our ability to fight infections.

This same genetic analysis confirmed that the Neanderthal skin colour was light, but probably as variable as we see in modern Europeans. Their eye colour ranged from brown to blue, or blue-green or possibly hazel. Those of us who have inherited the lighter skins of Western Europe may have inherited part of our appearance from our Neanderthal distant grandparents. We may also have inherited our inclination to red hair and freckled skin, with the tendency to sunburn that goes with it. Moreover, as a report in the *Sunday Times* newspaper explained, part of our Neanderthal legacy may include a particular variant of the Major Histocompatibility Complex – the part of our genome that defines self and deals with foreign invaders – that increases the genetic risk of specific diseases such as type 2 diabetes, lupus and Crohn's disease.

On a lighter note, when Professor Stringer tested British comedian Bill Bailey and BBC science presenter Alice Roberts for the levels of their Neanderthal ancestry, he found that Bailey had inherited about 1.5 per cent of his genome from the Neanderthals, and Roberts had inherited 2.7 per cent. Stringer's own legacy lay in between, at 1.8 per cent. In his book, Pääbo would relate some comical extrapolations of the hybridisation going public, with men and women wondering if their Neanderthal inheritance explained the oddities they had observed in their own appearance, or behaviour, or the oddities they attributed to their spouses.

The deeper exploration of our hybrid origins and evolution is only just beginning. In particular scientists have only begun to examine the potential Neanderthal contribution to more subtle aspects of our physiology, immunological identity of self, our ability to counteract disease, including infections, and the differences, if any, between modern humans and Neanderthals in terms of intrinsic brain development, with its applications to cognition, creativity and the many social and cultural aspects of societal make-up.

*

A surprising discovery in the genome of the Altai female was a peculiarly low level of genetic diversity. She appeared to be the offspring of inbreeding parents who were related at the level of half-siblings. Occasional inbreeding is a feature of hunter-gatherer societies, including early modern humans. But this finding in the Altai genome raised an important question: were the Neanderthals, who appeared from the archaeological record to live in relatively small groups, with little evidence of wider mobility when compared to contemporaneous modern humans, more at risk of inbreeding than early modern humans? This will need to be explored further with high-quality sequencing of a wider range of Neanderthal genomes, but if inbreeding was significantly commoner in Neanderthal groups it might explain the impoverishment in genetic diversity seen in the

Altai genome. This in turn would have increased the risk of inborn errors of metabolism. An individual inheriting a single recessive gene from one of her parents would be protected from the disease if she inherited a normal variant of the gene from the other parent. If both parents were closely related, as in the Altai female's case, there would be a significantly greater risk that she would inherit the defective gene from both parents.

Another important revelation of the Altai genome was its suggestion of a surprisingly low population of Neanderthals at this late stage in their occupation of the Eurasian landmass. It raises a vitally important question: what were the relative population sizes of Neanderthals and modern humans at the time when the two populations met in Eurasia?

Of potential relevance is the climatic catastrophe known as the Last Glacial Maximum, which afflicted the ecology of the Neanderthals roughly 48,000 years ago, a freeze so severe that vast areas of the land were buried under glaciers miles deep and vast tracts of the northern Atlantic Ocean were similarly frozen over. There is evidence that this decimated the population of animals and people surviving throughout Eurasia. Could this explain the reduced numbers, and tendency to inbreeding, of the surviving Neanderthals at the time, a catastrophe that still afflicted their numbers some five thousand years later when our modern human ancestors arrived?

We are now in a position to take a more measured look at what might really have happened to the Neanderthals, who were more similar to ourselves than we had previously thought. Their populations may have been reduced to relatively small and infrequent hunter-gatherer bands by the time our ancestors met them. This makes all the more relevant Mellars' and French's suggestion that at the time of early co-habitation of Europe by the two peoples, our ancestors may have outnumbered the Neanderthals population by roughly ten to one. While we cannot rule out skirmishes, or even the extermination of some population groups, we might question how purposeful hostilities or massacres might have been in such circumstances. We know that the two populations interbred, perhaps with less inhibition and on a larger scale than some would like to imagine. Such interbreeding between a small and a much larger population is likely to result in the bulk of the smaller population being assimilated into the larger.

Could this then be the explanation for the mysterious fate of the Neanderthals?

It would certainly explain why we find none of the pathognomonic skeletal features in the Eurasian population within 10,000 to 15,000 years of modern human arrival – this is plenty of time for the Neanderthal cranial and skeletal features to be melded into and swallowed up within the rapidly expanding, and still evolving, larger population of early *Homo sapiens*. If this is what happened, the Neanderthals did not become extinct – or at least not in the way we previously imagined. They disappeared as a separate, distinguishable population but live on as an integral part of our own hereditary pedigree.

In April 2014, Paolo Villa, from the University of Colorado, and Wil Roebroeks from Leiden University in the Netherlands, teamed up to write an overview of what they termed 'the Modern Human Superiority Complex', by which they meant the overweening assumption by many scientists and public alike as to our superiority over the Neanderthal cave men and women. They concluded: 'This systematic review of the archaeological records of Neanderthals and their modern human contemporaries finds no support for such interpretations, as the Neanderthal archaeological record is not different enough to explain the demise in terms of inferiority in archaeologically visible domains.' Instead they proposed that complex processes of interbreeding and assimilation was indeed the more likely explanation for the disappearance of the Neanderthal skeletal and physical features from the fossil record.

*

It is an extraordinary story, brought about by extraordinary scientific breakthroughs in genetic research, but still there were more surprises to come during the fruitful period in which Pääbo's breakthrough was being applied to hitherto refractory fossil bones.

In July 2008, a Russian archaeologist, Alexander Tsybankov, was digging in the floor of the same cathedral-like cavern high up in the Altai Mountains of southern Siberia. Working through deposits dating back 30,000 to 50,000 years, Tsybankov discovered a fragment of a single finger bone, an unprepossessing chip from the tip of the little finger. When he showed his find to his boss, Anatoly Derevianko, the latter thought that the bone was probably modern human. This would explain how it came to be in the same deposit as some sophisticated artefacts, including a bracelet of polished green stone. But Neanderthal remains had previously been found in the same cave so Derevianko chopped the fragment of bone into two parts, putting the smaller of the two into an envelope and arranging for it to be hand delivered to Pääbo in Germany for genetic analysis. The tiny sliver of bone arrived just as Pääbo's team were about to complete the first draft sequence of the Neanderthal nuclear genome, and the team were already very busy, with a backlog of fossils waiting to be examined. The Russian bone went to the back of the queue.

It wasn't until late 2009 that Pääbo's colleague, Johannes Krause, assisted by Chinese graduate student Qiaomei Fu, found the time to perform a preliminary screen of the Altai bone's mitochondrial DNA. Minuscule as the fossil was, it proved to be incongruously rich in DNA and it was also relatively uncontaminated. And what they found was so startling that they were obliged to repeat the analysis. The strange findings were unchanged. An excited Krause picked up the phone and called Pääbo, who was attending a meeting at Cold Spring Harbor Laboratory in New York. Krause began by asking him: 'Are you sitting down?'

He wasn't sitting down.

'Maybe you had better find a chair?'

Pääbo would subsequently confess that he had indeed found himself a chair because he feared that something terrible had happened. The mitochondrial genome extracted by Fu was not that of a Neanderthal. It was so surprising that Krause had counter-checked it because he couldn't believe Fu's findings. He had then insisted on comparing it to all six versions of the Neanderthal mitochondrial genomes now filed in their records. Without question, it was not Neanderthal. No more was it the mitochondrial genome of any of the modern humans that had been sequenced from all around the world. Whereas the Neanderthal mitochondrial genome had differed from that of modern humans in 202 nucleo-tides, or Snips, this differed in 385. The staggering truth was that it wasn't like any mitochondrial genome they, or anybody else, had ever sequenced before.

The implications flashed through Pääbo's mind. Could it be that they had turned up a hitherto unknown species of human? Working on the basis of the Snips, if the Neanderthals had split from modern humans around 500,000 years ago, then this species must have split from a common ancestor maybe 800,000 years ago. Yet a member of this same species must have been alive in Siberia some 50,000 to 30,000 years ago? In Pääbo's recollection: 'My head was spinning.'

When he got back to the Max Planck Institute, some three days later, he talked over the findings with Krause. The sliver of finger bone was, it seemed, almost miraculously rich in DNA. For example, the very best source of Neanderthal DNA in the fossil bones they had tested had yielded 4 per cent, but the new bone yielded 70 per cent. Not only had it come from an amazing human source, it had experienced an equally amazing level of preservation. Nevertheless, Pääbo insisted that Krause and Fu repeat the analysis on what little was left of the tiny bone. The results were exactly the same. He emailed Anatoly Derevianko and they arranged to meet at the Institute of Archaeology and Ethnography in Akademgorodok, a Russian city that had been purpose-built for science by the Soviet regime in the 1950s. Pääbo arrived in deep winter, with an ambient temperature of 35 degrees below zero. He knew that the sliver of bone had been part of a slightly larger whole and he asked for the remainder to work on the nuclear genome. But they told him they had sent it to a colleague in America, who appeared to have lost it.

Chagrined, the German team returned to Leipzig, though they had succeeded in bringing back a most unusual tooth that the Russian archaeologists had recovered from the same cavern in the Altai. The tooth was unusually large in size and primitive in appearance when compared to that of a modern human or a Neanderthal.

On 10 April 2010, Pääbo's team and their Russian colleagues published their findings of the extraordinary mitochondrial genome in the Letters section of the journal, *Nature*. It was a unique paper, the first time an extinct human species had been discovered by genomic analysis of a fossil bone. In the scientists' opinion, the most likely explanation of the Altai finger bone is that it represents 'a hitherto unknown type of hominin' – one that, a very long time ago, shared a common ancestor with modern humans and Neanderthals. There were some 'exceptionally archaic features' to the mitochondrial DNA that tallied with the archaic nature of the two teeth. The differences between this new hominin and the other two were so marked that they considered formally announcing it as the discovery of a new species, but then – prudently, as it turned out – they changed their minds.

The cave where the sliver of bone was discovered had been inhabited by a hermit named Denis back in the eighteenth century. So they decided that they would label the newly discovered people the 'Denisovans'.

And still the litany of surprises was growing . . .

The tiny chip of bone was so rich in DNA that they managed to extract a high-quality nuclear genome from it. In December of the same year, Pääbo's team combined forces with David Reich, of Harvard Medical School, and colleagues from a large number of other institutes in America, Germany, Spain, China, Canada and Russia, to publish the Denisovan nuclear genome. They could now confirm that at the time when our modern human ancestors emerged in Africa, around 180,000 years ago, a number of cousin species shared the world with them. First they dealt with the modern human and Neanderthal split. They gauged, from the two different genomes, that the human lineage and Neanderthal lineages had separated into separate lines a little later than they had earlier surmised, between 270,000 and 440,000 years ago. This confirmed that the split was probably too recent for complete species separation, thus allowing interbreeding between the two emerging species with fertile hybrid offspring. Examination of the Denisovan genome and comparison with that of Neanderthals suggested that the Denisovans had also shared a common ancestor with the Neanderthals, but the split between these lineages had been much earlier than that of Neanderthals and modern humans, at roughly 640,000 years ago. The last shared ancestor of the Denisovan and human ancestral lineages was put even further back, at roughly 800,000 years ago.

The comparative genetics showed that the Denisovans were more closely related to the Neanderthals than they were to us, but not so close that they extensively mated with them. It also confirmed that they were not merely a subgroup of Neanderthals but a separate species very likely inhabiting a wide geographic area of Asia, and with an evolutionary history distinct from that of modern humans and Neanderthals. Some of the genetic sequences in the Denisovan genome looked so primitive that it made them wonder if the Denisovans had acquired these sequences through hybrid crossing with a more archaic species still. The latter was described as an as-yet-unidentified species, but the most likely candidate is fascinating: it could signify our first glimpse into the genome of the common ancestor of all three species – modern humans, Neanderthals and Denisovans – the amazing global traveller and pioneer of early humanity, *Homo erectus*.

Like the Neanderthals, the Denisovans had interbred with early modern humans, but where the Neanderthals had contributed to most European lineages, the Denisovans had only contributed to Asian lineages, and in particular to native peoples of Polynesia, Melanesia and Australia. The geneticists found that the Denisovans had contributed some 4 to 6 per cent of the genome to Melanesians who currently inhabit areas of southeast Asia, suggesting that the Denisovans inhabited a large geographic area in Asia long ago. The paper ended with an emerging picture of a distant period of human evolution, known as the Upper Pleistocene, in which 'gene flow among different hominin groups was common'. Not species genocide then, but mutual sharing of culture and genetic inheritance.

Within a few years of the extraction of the Denisovan nuclear genome, an international group of geneticists confirmed the advantage of hybrid genomic novelty in terms of survival when the going got tough. One of the most celebrated examples of such an adaptation in humans is the ability of Tibetans to survive extreme high altitude in their Himalayan homeland. Geneticists had already discovered that Tibetans possessed a unique 'hypoxia pathway gene', known as *EPAS1*, that lowered their haemoglobin levels in conditions of low oxygenation – the very opposite of what happens when a non-Tibetan is exposed to such a low-oxygen atmosphere. The normal thickening of blood under such conditions would put people at risk of life-threatening blood clots. Tibetans only shared the gene *EPAS1* with one other group of people – the Denisovans. In the researchers' opinion: 'Our findings illustrate that admixture with other hominin species has provided genetic information that helped humans to adapt to new environments.'

The surprises just kept on coming.

*

The Atapuerca Mountains, in the northeast Spanish province of Burgos, are honeycombed with caves containing hominin fossils and artefacts. One of these caves, known as the Sima de los Huesos – the 'pit of bones' – has yielded one of the greatest assemblage of hominin bones ever found, including the remains of at least 28 individuals that have been dated to more than 300,000 years old. The skeletons have some features that resemble those of the Neanderthals, but many of their features are more typical of the more archaic *Homo heidelbergensis*, thought by some to be the ancestor of Neanderthals, and by others to be the ancestor of both modern humans and Neanderthals. The promise of this archaeological treasure trove was heightened by the fact that these bones were remarkably well preserved, suggesting they might be a good source of archaic DNA. Spanish palaeontologists supplied Pääbo with a complete femur in remarkably good condition, and this was duly drilled to supply 1.95 grams of powdered bone. As earlier, the geneticists began by sequencing the mitochondrial DNA, to produce yet another round of astonishment when they revealed the results in a *Nature* article, published online in December 2013 and in printed form the following January.

The experts had anticipated a mitochondrial genome closely related to, and very likely ancestral to, the Neanderthal sequences. But what they found was a genome that was closer to that of the Denisovans than to either the Neanderthals or to modern humans. They had provoked yet another mystery that appeared to shake up our prior ideas about our human origins.

In a covering article in *Nature*, the authors confessed that they were now scratching their heads to explain the surprising discovery. Everyone seemed to have their own different ideas as to the explanation. As Clive Finlayson, an archaeologist at the Gibraltar Museum, commented, the findings were actually somewhat 'sobering and refreshing'. Too many ideas about human evolution had been derived from limited samples and overblown notions. From now on the real truth would be revealed by the genetics, which, in Finlayson's words, 'doesn't lie'. In Pääbo's admission, he was as bemused as everybody else. 'My hope is that eventually we will bring not turmoil but clarity to the situation.'

What a refreshing light palaeogenetics is shining on the real history of our distant ancestors!

While we cannot rule out skirmishes, even limited violence, between the different species and populations, it would appear that the different evolving species of humans were not intent on exterminating their evolutionary rivals. They surely came across one another from time to time – they may even have lived cheek by jowl in some geographic areas – when, to judge from recognisable human behaviour, they would probably have been intensely curious about one another. They must have recognised their common humanity, held conversations with one another, looked at each other's traditions. They may have learnt from one another, too, picking up ideas about hunting or foraging, or information on the manufacture of tools, how they operated as groups, or the traditions of family life, the rules of sexual partnership, how they took care of and educated children, how they decorated their bodies, manufactured their clothing and homes, how they worshipped, or grieved for and dealt with their dead.

This is what Pääbo and his colleagues want to know more about. It is also what you and I want to know more about – the real human history, the story that is forever retained within and being constantly added to anew in that mysterious world of our common human genome.

nineteen
What Makes You Unique

> *The preservation of favourable variations and the rejection*
> *of injurious variations, I call Natural Selection. Variations*
> *neither useful nor injurious would not be affected by natural*
> *selection, and would be left a fluctuating element ...*
>
> CHARLES DARWIN

In the celebrationary issue of *The Daily Telegraph*, on Monday 12 February 2001, Roger Highfield, the newspaper's science editor, forecast that cracking the mysterious code of the human genome made every human being special. He was absolutely right. Just how special is, of course, still in the process of being unravelled.

It will have long been apparent on this journey that we all share the genomic legacy of a fascinating and extraordinary evolutionary history. It is a history that straddles the very beginnings of life on Earth and what is turning out to be an epochal age, in which our species is beginning to explore the Universe beyond our ocean-girdled planet. We caught a snapshot of this history when, in 2001, the first draft of our human genome revealed that we share thousands of genes with many other forms of life, and not just the great apes or even the mammals, but with reptiles, fish, the fruit fly and the nematode worm. And it goes even deeper than this. My late friend and distinguished scientist Lynn Margulis showed us how we have inherited so much of our distant prehistory, and much of our fundamental internal chemistry, from the bacterial stage of life – in the jargon, the Proterozoic stage – that pioneered many of the genes and metabolic pathways that life depends on today. And, chapter by chapter, we have discovered how all four of the mechanisms that give rise to hereditary change, those same mechanisms I grouped together under the umbrella concept of 'genomic creativity', have provided the 'variation' necessary for Darwin's pioneering idea of evolution to mould our wonderful human genome.

We have seen how the symbiotic union of the genomes of former parasitic microbes with the genomes of our ancestors has contributed to this inexorable evolution, from the capture of the energy of sunlight by cyanobacteria, and the production of high-energy oxygen as a by-product, to the respiration of the oxygen by the bacterial forerunners of the mitochondria that now add a second genome to our living cells, as well as the invasion of the endogenous retroviruses that are still changing the way our genome works. We have seen how, as the genome became more and more complex, powerful systems of epigenetic regulation have become intricately involved with the bureaucratic systems of governance of genes and many other aspects of the genome. Some scientists would now regard the genes as hardware and the regulatory systems as software, with the implications that where the hardware is for ordinary purposes fixed, the software is capable of changing in relation to signals from the environment in every individual human being. And we have seen how sexual crosses between cousin species have added major injections of genetic diversity to our evolving ancestral genome.

All of this may appear a confusion of competing forces – and so it would be if evolution was random, but thanks to Darwin we realise that it is not random. There is an additional powerful editorial force – Darwin's brilliant concept of natural selection – which selects for those changes in heredity that enhance survival and selects against those that threaten survival. Survival, and through it reproduction, governs every mechanism that contributes to what might appear a confusion of competing forces. And yes, built into this history of our constantly evolving genome is the inevitability that every one of us is genomically unique.

We are unique, to begin with, because each of us, other than genetic twins, inherits a randomised mixture of the genomes of two different individuals – our parents. The mixing is inherent in the way the germ cells are created within the ovaries and testes of our mothers and fathers. It is brought about during the process known as 'meiosis', when the chromosomes line up in parallel to one another and then the similar chromosomes break up into fragments and swap these matching fragments with one another. This process of sexual homologous recombination explains why brother is not identical to brother and sister to sister, even though they share the same parents. Only identical, or so-called 'monozygotic', twins share identical genes, because they develop from the same fertilised egg. But now, following our exploration of epigenetics, we know that even identical twins are already developing differences in their epigenetic systems of regulatory control by the time they are born. And if we looked really closely at their genomes throughout life, we would discover that they become increasingly different, because their epigenetic systems have been responding to different environmental stimuli.

A key region in making every human being unique is a portion we have repeatedly visited on our journey, the major histocompatibility complex, or MHC. Located on chromosome 6, this contains more than a hundred protein-coding genes, and it determines our immune defences as well as our antigenic identity, for example when it comes to blood transfusions and organ transplants. No part of our genome so tellingly defines us as 'self'. This personal genetic identity begins within the developing embryo in the mother's womb and it continues to update itself, through interaction with invading microbes, throughout all of our lifetime. It is through some damage or aberration of this complex recognition of self that autoimmune diseases such as rheumatoid arthritis, lupus and juvenile onset diabetes arise.

We have seen how tiny errors are made every time the DNA of our genome is copied to give rise to the ovum and sperm – these mutations in parts of our DNA that are of no consequence to natural selection give rise to the Snips, haplotypes and haplogroups that enable genetic historians to trace origins and movements of historic populations.

Thus at conception we share roughly half our genes, including those vast numbers of single nucleotide polymorphisms, or Snips, with each parent. We also share roughly half of our genes and Snips with our siblings. If we have identical twins, we begin our embryological development by sharing all of our genes and Snips with our twins. In similar fashion we share a quarter of the same with our grandparents, an eighth with our great-grandparents, and so on back in time. But there is already potential for change even in this seemingly well-ordered system. So vast is the genome that there is a measurable potential for small mistakes during its copying. And those mistakes guarantee that tiny parts of us will be different even from the genetic sequences we would have expected to inherit from the genomes of our parents.

Whole genome sequencing has now established the mutation frequency for whole genomes. From one generation to the next – in other words from parents to child – there are, on average, about 70 new mutations. The vast majority of these are not located in the 1.5 per cent protein-coding fraction of the genome, where the average is a single mutation for every three parent-to-child generations. Instead, the majority are to be found in the viral and epigenetic regulatory regions. We shall return to this shortly, but I would like to continue to focus on mutational change. As an integral part of this predictable mutational change, you and I can expect to have Snips unique to our genomes. Something closely related to this has contributed to the uniqueness of individual genomes that is essential to what we know as DNA fingerprinting.

*

We are familiar with DNA fingerprinting as a means of determining family relationships, for example in paternity testing, or identifying the perpetrator of a crime. Up until the 1980s, accurate forensic identification had largely relied on fingerprinting, but in many crimes there was no fingerprint evidence. DNA profiling offered the same accuracy of individual identification, whether from saliva, or a spot of blood or semen, or a sample of tissue of any kind, including bone. But first there was a workaday methodological problem that needed to be overcome; a busy forensic service could not be expected to screen an entire human genome, with all of its 6.4 billion nucleotides, in the search for the elusive evidence of individual peculiarities. What was needed was a simple and reliable system of automated screening capable of spotting differences between individuals to the same high fidelity as fingerprinting. In 1985, a Leicester-based British geneticist called Alec Jeffreys provided this.

Jeffreys made his discovery by accident when exploring differences in DNA sequences between individual family members of one of his laboratory technicians. He had been examining odd-looking DNA sequences from the 'repeat' sections of the genome – those huge chunks of virus-related sequences that were scattered widely throughout the chromosomes. Here and there he observed regions of DNA containing repeats of the same handful of nucleotide letters. These so-called repeats were hardly uncommon in the human genome, but in certain locations within the chromosomes the actual number of them seemed to vary from one individual to another.

If we were to pay one such region a visit on our magical steam train, we would find ourselves hopping down to walk along a highlighted section of track, noting that the sequence began with, say, four sleepers, that read perhaps T, C, A and G. As we continue to perambulate the track, we find the same sequence, TCAG, repeating itself, possibly three times. Since these repeats occur in tandem along the track, they are called 'tandem repeats'. As far as Jeffreys could see, they served no purpose in the sense of the DNA translating to protein – or, in these more enlightened times, to genomic regulation, genetic or epigenetic. These were typical of the sort of sequence that would be ignored by natural selection because they would make no difference to individual survival or reproductive capability. So if we were to examine many different individuals within a population, through chance alone the numbers of tandem repeats at these sites would be very variable. They might show closer than average similarity among siblings, other than identical twins, but even siblings would show some differences as a result of sexual homologous recombination. In the jargon, these sites were locations of 'variable number of tandem repeats', or VNTR.

What Jeffreys did next was to develop a simple methodology based on the numbers of repeats at ten different VNTR locations scattered throughout the chromosomes. Why ten? We might indulge in the same simple mathematics we used earlier to determine how many nucleotides we needed in an overlap between fragments of chromosomes for it to be significant beyond reasonable doubt. Ten loci, say with variation from 1 to 4 repeats, proved to be more than enough to identify an individual beyond phenomenal levels of reasonable doubt. Jeffreys then added a simple genetic test that would determine whether the forensic evidence came from a male or a female. As we might have expected, the effectiveness of the actual genetic screening is greatly improved by using PCR, which needs only trace amounts of an individual's DNA to find a match. Thus forensic scientists were given an incredibly accurate new tool of individual identification, based on the very fact that every human being really is genomically unique; and that evidence could be gleaned from a trace of blood, or body fluids, from a single hair, or the cells shed from skin, indeed from a very wide variety of personal identification left behind at a great many different crime scenes. It just remained to be demonstrated that the new genetic methodology would prove to be every bit as helpful as traditional fingerprinting.

One of the earliest applications of genetic fingerprinting was in the search in the county of Leicestershire for a rapist killer of two teenage girls. Not only did this process discover the real murderer, it also exonerated an innocent man who until then had been considered the prime suspect. Since then Jeffreys' methodology has been taken up by forensic laboratories around the world, helping to solve a vast array of family pedigree genetic enquiries as well as criminal cases. But we should not confuse DNA fingerprinting with the complete DNA sequencing of an individual human genome. This remains a formidable undertaking, although it is much easier to conduct these days, with high-throughput computer-assisted sequencing machines. Whole genome sequencing is becoming increasingly common, for various purposes, and this has highlighted how unique each and every individual human being really is at levels that go far beyond variable tandem repeats.

*

A single endogenous retrovirus insert, or locus, is roughly 10,000 nucleotides long. People hailing from Africa or the Near East are much more likely to contain the loci of HERV-113 and HERV-115 in their genomes than people originating in Western Europe, or Asia. Meanwhile, those same Western Europeans are far more likely to contain tracts of DNA of Neanderthal origins than people hailing from sub-Saharan Africa. Asians and Polynesians are also likely to contain some Neanderthal DNA, and to contain even more Denisovan DNA, amounting to as much as one-sixteenth of all their genome. How much information will the more detailed exploration of these differences provide of our human history of origins and migrations that was previously thought lost to the dust and fossils of prehistory? We are only beginning to explore the implications of these two major hybridisation events. Yet such differences do not feed into the slanted viewpoints of racists. Rather, they confirm and extend what the earliest geneticists, such as Luigi Luca Cavalli-Sforza, were at pains to emphasise and celebrate: our oneness not only as a species but as a human family.

Whole genomic sequencing must, by definition, include both the mitochondrial and the nuclear genome. This has led to surprising differences when we screen populations. We have already seen significant differences between males and females when we conduct haplogroup screening of European populations, with females showing a much greater homogeneity throughout all populations when compared to males. Sexual differences in haplogroup movements are also seen in much more recent population screenings, for example in the peopling of the various nations among the British Isles. Whole genomic sequencing may help to clarify what this might mean. As we saw with the Neanderthals, screening of the mitochondrial genome, or perhaps the similarly limited screening of Y-chromosome sequences, may actually be giving us subtly different information from the screening of the entire nuclear genome. How interesting if this translates to real differences in the prehistoric movements of the two sexes within ancient societies!

Some of the first whole personal genomes to be sequenced included J. Craig Venter, the entrepreneurial scientist who led the commercial group to the first draft sequence of 2001, and James Watson, the co-discoverer of DNA. When Korean researchers compared their genomes to those of a Han Chinese, a Yoruban Nigerian, a female leukaemia patient and a Korean-originated scientist, the researchers were astonished to discover that the two Americans had more sequences in common with the Korean than with one another. I should add that what the Korean researchers actually compared were not whole genome DNA sequences but whole genome patterns of Snips, when they discovered some 420,083 novel, single-nucleotide polymorphisms previously unknown to the Snip database as well as the startling similarities and differences mentioned above.

Another interesting personal example was the 2008 sequencing of the mitochondrial genome of Ötzi, the Tyrolean Iceman, whose 5,300-year-old mummified remains had been discovered with the thawing of an Alpine glacier. This revealed that he belonged to a branch of the mitochondrial haplotype K1 that had not been identified in European populations up to this point. But when, in 2011, this was followed by Snip analysis of his nuclear genome, it showed a recent common ancestry with the inhabitants of the Tyrrhenian Sea, which is part of the Mediterranean immediately west of Italy, including the coast of Tuscany and the islands of Corsica and Sardinia. The report, by Keller and colleagues, suggested that he had brown eyes, belonged to blood group O, was lactose intolerant and had a genetic predisposition to coronary heart disease.

When Pääbo and his colleagues were screening archaic genomes they made some startling additional discoveries. The different sections of our chromosomes have different palaeontological origins in terms of people and time. It would appear that when we visit these different regions of the genomic landscape in our magical train, we really are glimpsing something of those different ancient peoples in their prehistoric worlds. A related discovery was that while some portions of our genome have a very modern evolution, other portions appear to have stayed the same for millions of years. When Pääbo and his colleagues calculated back to the likely date of the common ancestor of the 'reference modern genome' and the Neanderthals, the results suggested an LCA some 830,000 years ago. But when they dated the last common ancestor of the same reference modern genome and the San people in Africa, they came up with a date of 700,000 years. Even stranger, when they dated last common ancestors for specific present-day DNA regions of chromosomes from people from different geographic locations over the Earth, they found some regions where they shared a common ancestor just 30,000 or so years ago, but other regions that suggested a last common ancestor 1.5 million years ago – the time of early *Homo erectus*. In Pääbo's own words: 'If somebody could take a walk down one of my chromosomes and compare it to both a Neanderthal and the reader of this book, that chromosomal pedestrian would find that sometimes I would be more similar to the Neanderthal than to the reader, sometimes the reader would be more similar to the Neanderthal, and sometimes the reader and I would be more similar.'

For Pääbo's walk, my reader and I might substitute a train ride, and I think we might share a smile for a moment.

It is strange to consider that different regions of our genome have different evolutionary origins. But perhaps we shouldn't be so overly surprised by it when we realise that we share a thousand or more genes with worms and fruit flies and, in the opinion of my friend and colleague, the eminent evolutionary virologist, Luis Villarreal, we have inherited key DNA and RNA management genes from viral lineages that date to billions of years ago. Sometimes an evolved genetic or genomic system simply works so well that the passage of time cannot improve on it, over millions, or even billions, of years. Yet this was exactly the point made by Darwin, in his ruminations before the publication of his iconoclastic book. Our common ancestors, generation by generation, species to families, and back even beyond Phyla and Kingdoms, go back to the very origins of life on Earth.

twenty
The Fifth Element

*I wanted to take us into a new era of biology by generating
a new life form that was described and driven only by DNA
information that had been created in the laboratory.*

J. CRAIG VENTER, *LIFE AT THE SPEED OF LIGHT*

The philosophers among the ancients believed that matter, and thus
the Earth, was made up of four elements: Earth, Air, Fire and Water.
They further believed that the stars in the fiery heavens were made
of a fifth and more wondrous element which was infused with
the celestial power over life. These metaphysical elements are not
the same as the chemical concepts of elements today, which are the
building blocks of molecules, but they do bear an ideational comparison since, in a more holistic sense, the elements of the ancients
were, in their imaginations, the building blocks of worlds. And we
might continue the comparison in extrapolating the fifth metaphysical element to the remarkable sibling molecules, DNA and
RNA, which make possible the evolution, heredity and development
of life. How daunting then to open our own imaginations wider still,
to acknowledge that, for good or for bad, that quasi-miraculous fifth
element has now fallen into the ambitious grasp of humankind.

Tinkering with the processes of life is not new to humanity. As long ago as the Stone Age farmers learnt how to select the seed grains of wheat and other cereal crops to get fatter, more nutritious kernels. Today virtually all of the seed grains 'husbanded' by farmers are the results of hybridisation – that same evolutionary mechanism of sexual crossing between different species. Humans have been learning from, and also intruding into, the secrets of nature for a very long time, but only very recently could we be said to have added a fifth to the hitherto four natural mechanisms of genomic creativity – those same mechanisms that natural selection relies on to enable the evolution of life on Earth. That fifth mechanism is the calculated genetic engineering of living genomes.

Where in the past any human-induced genetic modifications were brought about as the accidental effects of breeding farmed animals, pets and crops; now, thanks to the golden age of genetics, we are poised to take the reins of deliberate genetic and epigenetic control. This is not some fearful, or wonderful, thing that will eventually come to be; it has already been happening for a generation in terms of the genetic engineering of animals and plants. If to date it has not intruded into the human genome, I'm afraid that logically it seems merely a matter of time before it begins to be applied to humans.

In the initial media response to the 2001 publications of the draft genome we witnessed that it heralded a new way of looking at ourselves. What else have we been doing in this book other than looking at ourselves anew? It is difficult to consider such possibilities dispassionately. Yet it would appear timely to do so. Scientists, including molecular geneticists, are neither amoral nor unethical. The obvious applications of the dawning golden age of 'creative genetics' – or, taking on board the dawning importance of epigenetic regulation, should we call it 'creative genomics'? – will be for the potential good of humanity, in terms of medicine, for improved provision of food and, less explicitly, as part of the ongoing exploration of the wonders of nature.

What could be more justified than understanding the genetic basis of disease, so we can make use of such understanding to treat affected individuals as well as preventing disease in future generations? These twin aims have already begun and are rapidly expanding in terms of preimplantation genetic diagnosis and the selection of healthy embryos. Some members of society will have ethical or religious objections to this. Pioneers of molecular genetics and 'recombinant DNA', such as James Watson, Sydney Brenner and Paul Berg, pointed out that the prudent way to encompass such concerns is to ensure that non-scientists understand and are thus 'intelligently aware' of such scientific enterprise and ambition so that the safety, moral and ethical implications are routinely taken into consideration.

A good deal of the modern extrapolation of genetics and genomics does not involve worrisome genetic engineering. Much of the pharmaceutical research into epigenetics, including non-coding RNAs, is aimed at medical therapies that change the epigenetic control for the better. This is already the basis of lines of research into cancer therapy. I can predict that it will also become the basis of many lines of research into autoimmune diseases.

As we have seen, our physical and mental health is closely linked to genetic as well as environmental factors. Genetic and epigenetic differences between individuals may determine the potential for addiction to drugs and alcohol. A similar individual variation may be important in our predisposition to many different diseases. This is ushering in new fields of investigation and prediction of disease, such as the closely related 'personal genomics' and 'predictive medicine'. Personal genomics – also referred to as 'integrated personal omics profiling' – is an ambitious programme of research aimed at providing a dynamic assessment of the physiology and health of an individual over time. One such investigation is the brain-child of Michael Snyder, Professor of Genetics at Stanford University, in which volunteers are subjected to genomic, transcriptomic and proteomic high-throughput readouts, combined with screening of the individual's metabolic state, and changes in auto-antibody profiles. The idea is to spot key changes in and interactions between genome, epigenome and internal physiology during normal health and in the lead-up to disease.

Something similar is happening in other countries. In the UK, between 2006 and 2010 a registered charity called UK Biobank recruited 500,000 people aged between 40 and 69 years to undergo medical examinations and to donate blood for DNA, as well as urine and saliva samples for future analysis. The aim is to create a data bank that will improve our ability to prevent, diagnose and treat a wide range of serious and life-threatening illnesses – including cancer, heart diseases, stroke, diabetes, arthritis, osteoporosis, eye disorders, depression and forms of dementia. In 2005, in the United States, Dr George M. Church announced the creation of the Personal Genome Project, aimed at recruiting 100,000 volunteers from the US, Canada and the UK who would be agreeable to having their entire genomes sequenced and stored. This large collection of 'genotypes', or full DNA sequence of all 46 chromosomes, would be published, along with extensive information about the medical records, various physical measurements, MRI images, and so on, so that researchers will be able to study the links between genotype, the environment and so-called phenotype – the physical make-up and progress of the volunteers. Not only will this help to plot genetic links to disease, researchers are planning to examine the reaction of society, including insurers and employers, to such extrapolations from the genotype to future health predictions. Despite the potential for discrimination, it seems that recruitment has been very successful. It seems likely that similar wide-ranging genetic and epigenetic screening programmes will be conducted in many other countries.

In time such personal genome projects may enable predictive medicine, which is based on the notion that, by predicting the probability of serious disease from an individual's genome, this will allow active measures to reduce the risk of disease in the future. Another way in which this might prove useful would be to predict the likelihood of iatrogenic disease – disease caused by side-effects of medical therapy. In a recent study of adverse drug reactions involving 5,118 children admitted for both medical and surgical therapy to a UK-based hospital, 17.7 per cent experienced at least one adverse drug reaction. The authors thought it likely that the actual incidence of side-effects may have been higher because they excluded 'possible' but unproven cases. Opiates and anaesthetic drugs accounted for more than 50 per cent of the reactions and 0.9 per cent caused permanent harm or required admission to a higher level of care. It is important to grasp that many such side-effects were not life-threatening, for example vomiting after a general anaesthetic, but these experiences will have been frightening and memorable for children and would be better avoided, if possible. Another obvious potential for troublesome or even life-threatening side-effects is the long-term treatment of a very wide variety of diseases, both under hospital and general practitioner management. Some of the most serious side-effects may become predictable, and thus preventable by deciding from a choice of medication at the outset of therapy through modern 'omics' investigation.

Members of the public are also doing things for themselves. More and more people are paying to have their personal genomes sequenced, some merely through curiosity about their genetic background, others because they want to know about their own genetic predisposition to disease. For example, a woman who fears, from her family history, that she might be more susceptible to breast or ovarian cancer might want to know if she is carrying specific genes that'would increase her risk of these diseases, such as BRCA1 and BRCA2. This might allow her, in consultation with her doctors, to plan a course of action that would mitigate her risk.

All such investigations, as well as the therapeutic options that come from them, might invoke ethical, moral or religious dilemmas. We live in a rapidly changing world in which complex personal and social issues are being brought into question in ways our parents and grandparents would never have imagined possible, with increasing need for genetic counselling, genomic prediction and, perhaps soon, the potential for genetic engineering.

Even today some folk are worried that this growing understanding, and manipulation, of genetics will lead to a new potential for eugenics. Already some people might argue that preimplantation genetic diagnosis, with rejection of genetically compromised embryos, is an unacceptable form of eugenics, even though most affected families would probably see it as the preferred option in very painful and difficult circumstances. A commercially-run clinic in California is already providing to would-be parents a pre-determined sex of child. What else does the future hold? Is it going to become possible to genetically manipulate embryos to change their physical appearance, their stature, their athleticism, their intelligence? Will future generations of parents or governments instruct scientists to make use of DNA technology to breed what they regard as desirable genetic and epigenetic breeds of children?

*

I set out to write this book from the premise that it would attempt to provide a non-scientific reader with a basic understanding of how his or her own genome works. I can only hope that I have succeeded in that aim. The very notion that we might understand the evolution, structural make-up and detailed function of the genomes that code for life, including our own human genome, is of epochal importance not only for science but also for all of us. I hope that it has become clear that such understanding is important, since it must be for society in general, and not scientists alone, to decide where we go from here. Natural selection, nature's powerful force that decides what genetic novelty will move through a population to change the species gene pool, does not aim for any kind of perfection. As Darwin himself patiently explained, it is determined by one thing: survival, or failure of survival, of individuals, and implicit in this is the fate as to whether or not they reproduce offspring and thus contribute to the species gene pool. There is no higher objective involved in the moral, philosophical or religious sense – no forward planning whatsoever in the way human reason and ambition might conceive. But our capacity to alter genomes at our own whim changes that. Genetic engineering, if it is introduced into the human genome, will inject exactly such reasoned forward planning. This has important implications; the potential for the treatment and prevention of serious disease will obviously benefit society, but there will be other potentials that some might view as dangers. The moral and ethical implications are thus important. It is no exaggeration to say that what was previously the domain of science fiction is now increasingly science fact.

Genetic engineering of crops and farm animals began in the 1970s. From the beginning this process encountered societal resistance, some protests being more emotive than rational. But scientists, and governing bodies, were persuaded by the potential for benefit – purportedly the feeding of the hungry in parts of the world stricken by adverse climates and ecologies. Critics saw the potential for harm through GM-modified genes 'escaping' from the genetically modified fields to enter the surrounding ecology in unwanted ways. The crossing of genes from one species to another is called 'horizontal gene transfer'. As we have seen, genetic symbiosis, involving bacteria and viruses, and hybridisation events are potent examples of this crossing of evolutionary boundaries in nature.

In 1976 the US National Institutes of Health set up an advisory committee to advise on the putative dangers of 'recombinant-DNA', and this was followed by 'complex but relaxed' regulation from offices including the United States Department of Agriculture (USDA), the Environmental Protection Agency (EPA), and the Food and Drug Administration (FDA). This led to the establishment of a committee under the aegis of the Office of Science and Technology, which approved GM plants under the continuing regulation and control of the various regulatory bodies. In 2000 the Cartagena Protocol on Biosafety came into being as an international treaty to govern the transfer, handling and use of genetically modified organisms. One hundred and fifty-seven countries are members of this protocol, which is seen as a de facto trade agreement. GM crops will, usually, have built-in modifications aimed at preventing sexual crosses with non-GM crops. They also have 'traceability' built into their genomes, which would enable geneticists to discover the source of origin if GM-modified genes escape into the environment. In 2010, a study by US scientists showed that about 83 per cent of wild canola plants in the hinterland of GM crops contained genetically modified resistance genes. While the scientists involved in GM research and agriculture now saw no significant risk to the environment or humans from this established escape, those opposed to genetic engineering remain unconvinced.

A European deal in June 2014 opened up the possibility of

GM crops being grown commercially in any component state whose authorities decided to sanction this. The deal was supported by all of the member states other than Belgium and Luxembourg. Countries like France, which is said to oppose GM crops, will be free to eschew them, meanwhile England will be free to introduce them, even though other parts of the UK, including Scotland and Wales, may decide to oppose them. It is too early to speculate who is prudent and who is not.

The potential for future genetic modification of the human genome is even more likely to excite controversy and debate.

Most doctors would probably favour modification of the genome of people who are at high risk of genetically provoked serious, or potentially fatal, diseases – if and when such modification becomes available, and safe. Who wouldn't wish to save young women the risk of developing breast or ovarian cancer, or a child from cystic fibrosis, haemophilia, or Huntington's disease? But once the technology to modify the human genome becomes more readily available, how far will the applications extend? We began this journey with the aim of confronting the mysteries of the human genome, but now this very trail, and the solutions it has provided, may have opened a Pandora's box for future scientists, and society more generally.

Meanwhile, what of nature? Nature has no compunctions about changing genomes. So the question that is increasingly asked of scientists is as follows: is the human genome still naturally evolving today?

*

Our modern human history has been accompanied by dramatic changes in environment and lifestyle. We have been bombarded by lethal infectious diseases, including malaria, tuberculosis, yellow fever, pneumococcal pneumonia, meningococcal meningitis, whooping cough, measles, poliomyelitis, diphtheria. Many of these swept through populations on a regular basis, as well as notorious everyday bugs such as the staphylococcus and streptococcus that cause boils, cellulitis, rheumatic and scarlet fever, and bone and dental abscesses. In my lifetime I have treated human beings suffering from many of these illnesses. Susceptibility to diseases is one of the most powerful of external pressures for adaptive genomic change, most particularly affecting the evolution of the Major Histocompatibility Complex as well as the epigenetic portions of the genome. Meanwhile the lingering presence of resident retroviruses, and the hybridisation-derived Neanderthal and Denisovan genomic introgressions will inevitably be still working at the level of the species gene pool.

In 2006, Voight and colleagues from the Department of Human Genetics at the University of Chicago developed a new analytical method for scanning for Snips in whole genome surveys that was capable of searching for recent evolutionary pressures. In three broad geographic populations – east Asians, northern and western Europeans and Africans based on Yorubans from Ibadan, Nigeria – Voight and his team discovered widespread signals that denoted recent evolutionary change. These included genes related to malaria, lactose sensitivity, salt sensitivity in relation to climate, and genes involved in brain development. They also discovered signals of selective sweeps – so-called 'genetic bottlenecks' – that appeared to be still in progress and were presumably related to disease liability.

So, dear me – no! I do not imagine for a moment that we humans have stopped evolving.

Evolution is intrinsic to life. New viral plagues such as AIDS, hepatitis A, B and C, are sweeping through us. Natural calamities, including some that are man-made, are threatening us. We should recall that responsiveness to the environment is part of the way in which the epigenetic system evolves. And that epigenetic system is akin to an exquisitely sensitive and constantly changing software that governs how the genetic hardware works. A more subtle pressure may be the massive increase in knowledge and lengthening period of education of our young, this, coupled with the dramatic changes we have witnessed in just the last two decades in terms of how modern society works: think of the intrusion of computerised machines, social media, the global village, all maximally impacting on our young – a stage in which the human physiology, and epigenome, are still developing. Can we doubt that such overwhelming change is not already affecting our human evolution? How likely is it that such huge changes in behaviour and systems of learning, brought about by the IT revolution, will affect future brain development?

Meanwhile there is another related development, a potential change that may be the most astonishing of all: this is the capability of future genetic engineers to create artificial life forms.

*

Craig Venter, the scientist who founded Celera Genomics, produced the first commercially funded draft of the human genome in 2001, his team inventing several important innovations and developing the concept of ESTs and shotgun sequencing along the way. In a blunt and highly-entertaining autobiography, Venter declares that science has always set out to master life: 'For centuries a principal goal of science has been, first, to understand life at its most basic level and, second, to learn to control it.' Venter anticipates a future in which scientists will engineer new life forms, as well as modify the human genome, to suit human and societal needs. He has already taken what he sees as the preliminary steps to do exactly that.

By any stretch of the imagination, Craig Venter is an interesting individual. A man who took little interest in his early schooling in Salt Lake City, preferring to spend his time surfing and boating, he would subsequently put this down to his personal attention deficit disorder, which he had to struggle to overcome. Though opposed to the Vietnam War, he was drafted, enlisting into the US Navy, where he worked as an intensive care assistant in a field hospital. While in Vietnam, he attempted suicide by swimming out into the ocean, then changed his mind when more than a mile out. His experiences persuaded him to consider a career in medicine, but he subsequently changed course to biomedical research. Aggressively ambitious by nature, Venter proved to have a profoundly innovative gift as a scientist, combined with an irrepressible entrepreneurial spirit. In 2007 and 2008 he was listed in *Time* magazine's 100 most influential people in the world. Two years later he was listed fourteenth in the *New Statesman*'s roll call of 'the World's 50 Most Influential Figures'.

Venter was sacked by Celera in 2002, a year after the human genome was made public, reputedly because of differences in opinion with the main investor. He is currently President of the J. Craig Venter Institute, which has two main fields of enterprise. The first of these is to pioneer a discipline he labels 'synthetic biology', in which he and his colleagues aim to produce artificially engineered organisms designed to serve human and societal needs. He first began to work towards this goal with a company, Synthetic Genomics, that he founded in the early 2000s. He moved on to explore the minimal genome necessity for cellular life before synthesising the bare minimum of the genome of one of the smallest living bacteria, *Mycoplasma genitalium*, which causes urethral infections in humans. In essence he reconstructed the minimal genome in stages, first in his computer and then in a laboratory synthesiser. Prior to this, the largest genomes ever artificially assembled in this way had been the much smaller genomes of viruses, the first being the polio virus assembled by Eckard Wimmer and his colleagues. The *Mycoplasma* genome was twenty times larger. Overcoming many obstacles, Venter's group succeeded in replacing the natural genome of a living bacterium with his own synthesised equivalent to create a living bacterial cell – a breakthrough he and his colleagues published in 2010. This paved the way to the deliberate creation of cellular life forms to order.

The mystery has not ended, though. The extraordinary story of exploration of our mysterious human genome has, as ever, thrown up a new raft of very important questions.

Is Venter right in suggesting that Science has always been determined not merely to understand life at its most basic level but also to learn how to control it? It's something one is obliged to think deeply about. I can't say that I am sure as to the answer, but I rather suspect that he is. Why then is this so? Is it because we humans are arrogant enough to just think that we can? Or is it because we think that there are important reasons why we should? If Venter is right, we have gone beyond the stage of musing about this question. It is much easier to genetically engineer a germ cell, or a newly fertilised embryo, than to modify the genome of a developed human being. We have already genetically engineered animals and plants in this way. In April 2015 the human embryo was deliberately engineered in a scientific experiment for the first time. I believe that this is as great a leap as the discovery of gravity by Newton, relativity by Einstein, and the extrapolation of Einstein's discovery to the atomic bomb. And just as with those epochal discoveries, it carries with it the potential for great good and great harm.

bibliography

Brenner, S. *My Life in Science*. Philadelphia: Biomed Central, 2001.

Bronowski, J., *The Identity of Man*. New York: Prometheus Books, 2002.

Cavalli-Sforza, L. L. *Genes, Peoples and Languages*. London: Penguin Books, 2001.

Crick, F. *What Mad Pursuit: A Personal View of Scientific Discovery*. New York: Basic Books, 1988.

Darwin, C. *The Origin of Species*. London: John Murray, 1859. Penguin Classics reprint, 1985.

Darwin, C. *The Descent of Man*. London: John Murray, 1870. Prometheus Books edition, 1998.

Dawkins, R. *The Selfish Gene*. Oxford: Oxford University Press, originally 1976, 1989 edition.

Dubos, R. J. *The Professor, the Institute, and DNA*. New York: The Rockefeller University Press, 1976.

Duncan, D. E. *Masterminds: Genius, DNA, and the Quest to Rewrite Life*. London: Harper Perennial, 2006.

Friedberg, E. *Sydney Brenner: A Biography*. New York: Cold Spring Harbor Press, 2010.

Hartl, D. L. & Jones, E. W. *Genetics: Analysis of Genes and Genomes*. London: Jones and Bartlett, 2000.

Huxley, L. *Darwiniana: Essays by Thomas H. Huxley*. London: Macmillan and Co., 1893.

Huxley, J. M. *Evolution: The Modern Synthesis*. London: George Allen & Unwin Ltd, 1942.

Jablonka, E. & Lamb, J. M. *Epigenetic Inheritance and Evolution: The Lamarckian Dimension*. Oxford: Oxford University Press, paperback edition, 1999.

Judson, H. F. *The Eighth Day of Creation*. London: Penguin Books, 1995.

Luria, S. E. *Life: The Unfinished Experiment*. London: Souvenir Press, 1973.

Maddox, B. *Rosalind Franklin: The Dark Lady of DNA*. London: HarperCollins, paperback edition, 2003.

Margulis, L. *Origin of Eukaryotic Cells*. New Haven: Yale University Press, 1970.

Olby, R. *The Path to the Double Helix: The Discovery of DNA*. New York: Dover Publications, 1994.

Pääbo, S. *Neanderthal Man: In Search of Lost Genomes*. New York: Basic Books, 2014.

Pauling, L., *In His Own Words*. New York: Touchstone, 1995.

Ridley, M. *Francis Crick: Discoverer of the Genetic Code*. London: Harper Perennial, 2006.

Roberts, A. *The Incredible Human Journey: The Story of How We Colonised the Planet*. London: Bloomsbury, 2010.

Ryan, F. *Tuberculosis: The Greatest Story Never Told*. Bromsgrove: Swift Publishers, 1992. In the US, Ryan, F. *The Forgotten Plague*. New York: Little, Brown, 1993.

Ryan, F. *Darwin's Blind Spot*. New York: Houghton Mifflin, 2002.

Ryan, F. *Virolution*. London: Collins, 2009.

Ryan, F. *Metamorphosis: Unmasking the Mystery of How Life Transforms*. Oxford: Oneworld, 2011. In the US, Ryan, F. *The Mystery of Metamorphosis: A Scientific Detective Story*. White River Junction, Vermont: Chelsea Green, 2011.

Sayre, A. *Rosalind Franklin and DNA*. New York: Norton, paperback reissue, 2000.

Schrödinger, E. *What Is Life?* Cambridge: Cambridge University Press, paperback edition, 1962.

Shreeve, J. *The Genome War*. New York: Ballantine Books, paperback edition, 2005.

Smith, J. M. Szathmáry, E. *The Origins of Life: From the Birth of Life to the Origins of Language*. Oxford: Oxford University Press, 1999.

Stringer, C. & Gamble, C. *In Search of the Neanderthals*. London: Thames and Hudson, first paperback edition, 1994.

Venter, J. C. *Life at the Speed of Light*. London: Little, Brown, 2013.

Watson, J. D. *The Double Helix*. London: Weidenfield and Nicolson, 1968.

Wilkins, M. *The Third Man of the Double Helix*. Oxford: Oxford University Press, 2003.

chapter notes

Frontispiece quote: Pauling, L.: 25.

Introduction
Bronowski, J.: 4.

Chapter 1
Chapter head quote: Schrödinger, E.: 3.
The story of Dubos and his contribution to the antibiotic story: Ryan, F., 1992.
For more detail about Griffith, Arkwright, Dawson, Alloway, and so on: Dubos, R. J., 1976. Also Olby, R., 1994.
Avery and Dubos Cranberry Bog bacillus: Ryan, F., 1992.
Avery's paper: Avery, O. T., MacLeod, C. M. McCarty, M. Studies on the chemical nature of the substance inducing transformation of pneumococcal types. Induction of transformation by a desoxyribonucleic acid fraction isolated from pneumococcus type III. *J Exp Med* 1944; **79**: 137–58.
The letter from Avery to his brother is reproduced in Dubos, R. J., 1976.

Chapter 2
Chapter head quote: Deichmann, U. Early responses to Avery et al's paper on DNA as hereditary material. *Historical Studies in the Physical and Biological Sciences* 2004; **34**(2): 207–232.

On Avery not getting the Nobel Prize: Portugal, F. Oswald T. Avery: Nobel Laureate or noble luminary? *Perspectives in Biology and Medicine*, 2010; **53**(4): 558–70.

The Hershey and Chase paper: Hershey, A.D. & Chase, M. Independent functions of viral proteins and nucleic acid in growth of bacteriophage. *J Gen Physiol* 1952; **36**: 39–56.

The warmer reception of Hershey/Chase experiment, Olby, R.: 318.

The bombshell of Griffith's 1928 paper: Dubos, R. J.: 132–133.

For Alfred E Mirksky: Olby, R., 1994. Also Cohen, S. S. Alfred Ezra Mirsky: A biographical memoir. National Academy of Sciences.

Avery's response to Dubos, and Dubos' response to the death of Marie Louise: Ryan, F., 1992.

Chapter 3

Chapter head quote: Maddox, B.: 60–61.

Watson's early life and confessions of innate laziness: Watson, J. D.: 21.

Watson's early interest in phages: Judson, H. F.: 49.

Hershey 'Drinks whiskey but not tea': Judson, H. F.: 53.

Watson on Avery: Watson, J. D.: 13–14.

Chapter 4

Chapter head quote: BBC4 documentary, *The Code of Life*.

For more about Wilkins and Gosling: Wilkins, M., 2003.

Watson quotes: Watson, J. D., 1968.

'What Mad Pursuit': this became the title of Crick's autobiography, 1988.

Bragg was infuriated by the upstart junior: Crick, F., 1988.

Chapter 5

Chapter head quote: BBC4 documentary, *The Code of Life*.

Crick – 'Jim and I hit it off': Crick, F.: 64.

Mering as 'the archetypal seductive Frenchman': Maddox, B.: 96.

Franklin working better with Jewish male colleagues: ibid: 96.

Chargaff's scorn: Judson, H. F.: 142. See also Watson, J. D., 1968 and Crick, F., 1988.

Aaron Klug's comments about the terms of Franklin's appointment: Klug, A. The discovery of the DNA double helix. *J Mol Biol* 2004; **335**: 3–26.

Stokes figuring out the likely X-ray diffraction of a helical structure: Maddox, B.: 152.

Obituary for Franklin: Bernal, J. D. *Nature* 1958; **182**: 154.

Chargaff's paper: Chargaff, E. Chemical specificity of nucleic acids and mechanism of their enzymic degradation. *Experientia* 1950: **6**: 201–09.

Duncan's conversation with Watson: Duncan, D. E.: 169.

The three *Nature* papers on DNA:

Watson, J. D., Crick, F. H. C. Molecular structure of nucleic acids: a structure for deoxyribose nucleic acid. *Nature* 1953; **171**: 737–38.

Wilkins, M. H., Stokes, A. R., Wilson, H. R. Molecular structure of deoxypentose nucleic acids. *Nature* 1953; **171**: 738–40.

Franklin, R. E., Gosling, R. G. Molecular configuration in sodium thymonucleate. *Nature* 1953; **171**: 740–41.

Chapter 6

Chapter head quote: Judson, H. F.: 230.

For the chemical structure of DNA: Hartl & Jones, 2000.

Background of DNA to protein, see Judson, H. F., 1995; Olby, R., 1974.

For biographical information on Sydney Brenner: Brenner, S., 2001; Friedberg, E., 2010.

For different types of mutations: Hartl & Jones, 2000.

Chapter 7

Chapter head quote: Sanger, F. Sequences, sequences, and sequences. *Ann Rev Biochem* 1988; **57**: 1–28.

Brenner paper on *C. elegans*: Brenner, S. The genetics of *Caenorhabditis elegans*. *Genetics* 1973; **77**: 71–94.

Events of metamorphosis: Ryan, F., 2011.

Puberty and brain rewiring: Sisk, C. L. & Zehr, J. L. Pubertal hormones organise the adolescent brain and behaviour. *Frontiers in Neuroendocrinology* 2005; **26**: 163–74.

Chapter 8

Chapter head quote: Shreeve, J.: 236.

More about J. Craig Venter: Venter, J. C., 2013.

Roger Highfield quote: *The Daily Telegraph*: 4.

The *Nature* and *Science* papers:

International Human Genome Sequencing Consortium. Initial sequencing and analysis of the human genome. *Nature* 2001; **409**: 860–921.

Venter, J. C., Adams, M. D. et al. The sequence of the human genome. *Science* 2001; **291**: 1304–51.

Chapter 9

Chapter head quote: Huxley, T. H., 1893. Chapter V, Mr Darwin's Critics: 120.

Mutations and bat wings: Cooper, K. L. & Tabin, C. J. Understanding of bat wing evolution takes flight. *Genes and Development* 2008; **22**: 121–24.

Definition of genomic creativity: Ryan, F. P. Genomic creativity and natural selection: a modern synthesis. *Biol J Linnean Soc* 2006; **88**: 655–72.

Chapter 10

Chapter head quote: Margulis, L., 1970: Preface.

The Selfish Gene: Dawkins, R., 1976.

I have written a lot more about the history of symbiosis and its development in my book, *Darwin's Blind Spot*, 2002.

For Maynard Smith and symbiosis: Smith, J. M. & Szathmáry, E., 1999.

For Lynn Margulis, chloroplasts and mitochondria: Margulis, L., 1970.

Chapter 11

Chapter head quote: Wimmer, E. The test-tube synthesis of a chemical called poliovirus. *EMBO Reports* 2006; 7: special issue: S3–S9. The first sequencing of polio virus was published in Kitamura, N., Semler, B. L. et al. Primary structure, gene organisation and polypeptide expression of poliovirus RNA. *Nature* 1981; **291**: 547–53.

The viral components of the human genome: Ryan, F., 2009.

HIV-1 and human HLA-B types: Kiepiela, P., Leslie, A. J. et al. Dominant influence of HLA-B in mediating the potential co-evolution of HIV and HLA. *Nature* 2004; **432**: 769–74.

Screening for viral proteins in various tissues: Chen, F., Atterby, C. et al. Expression of HERV-R ERV3 encoded Env-protein in human tissues: introducing a novel protein-antibody-based proteomics. *J Roy Soc Med* 2013; **107**(1): 22–29.

Chapter 12

Chapter head quote: Jablonka, E. & Lamb, J.M.: vi.

C. elegans: Brenner, S., 2001.

Chapter 13

Chapter head quote: Brenner Nobel speech, 2002.

Chapter 14

Chapter head quote: Judson, H.F.: 10.

Clovis discovery: Rasmussen, M., Anzick, S. L. et al. The genome of a Late Pleistocene human from a Clovis burial site in western Montana. *Nature* 2014; **506**: 225–29.

Siberian child discovery: Raghaven, M., Skoglund, P. et al. Upper Palaeolithic Siberian genome reveals dual ancestry of Native Americans. *Nature* 2014; **505**: 87–91.

Richard III story: Ehrenberg, R. A king's final hours, told by his mortal remains. Sciencenews.org, 9 March 2013.

Chapter 15

Chapter head quote: Cavalli-Sforza, L.: 33.

First Wilson paper: Cann, R. L., Stoneking, M., Wilson, A. C. Mitochondrial DNA and human evolution. *Nature* 1987; **325**: 31–36.

For a more recent overview: Pakendorf, B. & Stoneking, M. Mitochondrial DNA and human evolution. *Ann Rev Genomics Hum Genet* 2005; **6**: 165–83.

Stringer's drawing attention to lack of agreement on date of Adam and Eve: Stringer, C. Out of Ethiopia. *Nature* 2003; **423**: 692–95.

Y–chromosome sequencing: Poznik, G. D., Henn, B. M. et al. Sequencing Y chromosome resolves discrepancy in time to common ancestor of males versus females. *Science* 2013; **341**: 562–65. Also Cruciani, F., Trombetta, B. et al. A revised root for the human Y chromosomal phylogenetic tree: the origin of patrilineal diversity in Africa. *Am J Hum Genetics* 2011; **88**: 814–18.

Mount Toba and Asian dispersal: Mellars, P., Gori, K. C. et al. Genetic and archaeological perspectives on the initial modern human colonization of southern Asia. *PNAS* 2013; **110**(3): 10699–704.

HERV-K106: Jha, A. R., Nixon, D. F. et al. Human endogenous

retrovirus K106 (HERV-K106) was infectious after the emergence of anatomically modern humans. *PLoS One* 2011; **6:** e20234.

Early modern human migration into Asia: Brown, P. Recent human evolution in East Asia and Australasia. *Phil Trans Roy Soc Lon Bio Sci* 1992; **337:** 235–42.

Aggressive symbiont: I have explained and defined what a viral aggressive symbiont comprises in three books, *Virus X, Darwin's Blind Spot,* and *Virolution,* as well as in many scientific papers.

Chapter 16

Chapter head quote: taken from Libby's Nobel Lecture.

The timing of early modern humans entering Asia: Mellars, P. Why did modern human populations disperse from Africa ca. 60,000 years ago? A new model. *PNAS* 2006; **103**(25): 9381–86.

Douka, K. Exploring 'the great wilderness of prehistory': the chronology of the Middle to the Upper Paleolithic transition in the Northern Levant. *Mitteilungen der Gesellschaft für Urgeschichte* 2013; **22:** 11–40.

Douka, K., Bergman, C. A. et al. Chronology of Ksar Akil (Lebanon) and implications for the colonization of Europe by anatomically modern humans. *PLoS One* 2013; 8(9): e72931: 1–10.

Mellars, P. & French, J. C. Tenfold population increase in Western Europe at the Neanderthal-to-modern human transition. *Science* 2011; **333:** 623–27.

Evidence for widespread recent evolutionary change in the human genome: Wang, E. T., Kidama, G. et al. Global landscape of recent inferred Darwinian selection for *Homo sapiens. PNAS* 2006; **103:** 135–40.

Chapter 17

Chapter head quote: from 'A conversation with Svante Pääbo'. Edge, 7 April 2009.

Sequencing genomes of animals and plants: Miller, W., Drautz, D. I. et al. Sequencing the nuclear genome of the extinct woolly mammoth. *Nature* 2008; **456**: 387–90. Dabney, J., Knapp, M. et al. Complete mitochondrial genome sequence of a Middle Pleistocene cave bear reconstructed from ultra-short DNA fragments. *PNAS* 2013; doi/10.1073/pnas.1314445110. Amemiya, C. T., Alföldi, J. et al. The African coelacanth genome provides insights into tetrapod evolution. *Nature* 2013; **496**: 311–16.

For Neanderthal fossils and appearances: Stringer, C. & Gamble, C., 1994.

Neanderthals and art: Abadía, O. M., & González Morales, M. Redefining Neanderthals and art: an alternative interpretation of the multiple species model for the origin of behavioural modernity. *Oxford J Archaeology* 2010; **29**(3): 229–43. See also, Zilhão, J. Symbolic use of marine shells and mineral pigments by Iberian Neanderthals. *PNAS* 2010; **107**: 1023–28.

Superiority complex with regard to Neanderthals: Villa, P. & Roebroeks, W. Neanderthal demise: an archaeological analysis of the modern human superiority complex. *PLoS One* 2014; **9** (4): e96424.

Chapter 18

Chapter head quote: from Zilhão, J. & Trinkaus, E. Eds. Portrait of the Artist as a Child: The Gravettian Human Skeleton from the Abrigo do Lagar Velho and its Archeological Context. *Trabalhos de Arquelogia* 2002; **22**: 9. ISBN 972-8662-07-6.

Neanderthal mitochondrial draft genome: Green, R. E., Malaspinas, A. S. et al. A complete Neanderthal mitochondrial genome sequence determined by high-throughput sequencing. *Cell* 2008; **134**: 416–26.

First draft Neanderthal nuclear genome: Green, R. E., Krause, J.

et al. A draft sequence of the Neanderthal Genome. *Science* 2010; **328**: 710–22.

ABO blood group: Lalueza-Fox, C. Genetic characterisation of the ABO blood group in Neandertals. *BMC Evolutionary Biology* 2008; **8**: 342.

Neanderthals and modern humans share the same variant of the language gene: Krause, J. et al. The derived FOXP2 variant of modern humans was shared with Neandertals. *Curr Biol* 2007; **17**: 1908–12.

Neanderthals found to have red hair and fair skin: Lalueza-Fox, C., Rompler, H. et al. A malocortin 1 receptor allele suggests varying pigmentation among Neanderthals. *Science* 2007; **318**: 1453–55.

Neanderthal ancestry in Asians: Wall, J. D., Yang, M. A. et al. Higher levels of Neanderthal ancestry in East Asians than in Europeans. *Genetics* 2013; **194**: 199–209.

The reappraisal of Neanderthal culture and society: Abadía, O. M. & Morales, R. G. Redefining Neanderthals and art: an alternative interpretation of the multiple species model for the origin of behavioural modernity. *Oxford J Archaeol* 2010; **29**(3): 229–43.

Teaching Neanderthal children: Spikins, P., Hitchens, G. et al. The cradle of thought: growth, learning, play and attachment in Neanderthal children. *Oxford J Archaeol* 2014; **33**(2): 111–34.

Compassion, culture of Neanderthals: Spikins, P. A., Rutherford, H. E. & Needham, A. P. From homininity to humanity: compassion from the earliest archaics to modern humans. *Time and Mind: The J of Archaeology, Consciousness and Culture* 2010; **3**(3): 303–26.

The complete Neanderthal genome from the Altai Cave: Prufer, K., Racimo, F. et al. The complete genome sequence of a Neanderthal from the Altai Mountains. *Nature* 2014; **505**: 43–49.

Genomic contribution of Neanderthals to present-day humans: Sankararaman, S., Mallick, S. et al. The genomic landscape of Neanderthal ancestry in present-day humans. *Nature* 2014: **507**: 354–57.

Neanderthal legacy in terms of disease: Leake, J. Neanderthals' revenge: the gift of deadly genes. *The Sunday Times*, 26.01.14: 19.

Disparity in human to Neanderthal populations when they met: Mellars, P. & French, J. C. Tenfold population increase in Western Europe at the Neanderthal-to-modern human transition. *Science* 2011; **333**: 623–27.

Discovery of the Denisovan genome. Marshall, M. Mystery Relations. *New Scientist*: 5 April 2014: 34–38.

Denisovan mitochondrial genome: Krause, J., Fu, Q. et al. The complete mitochondrial DNA genome of an unknown hominin from southern Siberia. *Nature* 2010; **464**: 894–97.

The Denisovan nuclear genome: Reich, D., Green, R. E. et al. Genetic history of an archaic hominin group from the Denisova cave in Siberia. *Nature* 2010; **468**: 1053–60.

Tibetan inheritance from the Denisovans: Huerta-Sánchez, E., Jin, X. et al. Altitude adaptations in Tibetans caused by introgression of Denisoval-like DNA. *Nature* 2014; doi:10.1038/nature13408.

DNA from Sima de los Huesos: The genome of a fossil from the Sima de los Huesos: Meyer, M., Fu, Q. et al. A mitochondrial sequence of a hominin from Sima de los Huesos. *Nature* 2014; **505**: 403–36.

Chapter 19

Chapter head quote: Darwin, C., 1859. Chapter IV: Natural Selection: 131.

Jeffreys' discovery of DNA fingerprinting: Jeffreys, A. J., Wilson,

V. et al. Hypervariable 'minisatellite' regions in human DNA. *Nature* 1984; **314**: 67–73.

Our oneness as a species: Cavalli-Sforza, L. L., 2001.

References to haplogroup differences between the sexes, in Europe and in Western Europe, including the British Isles: Wilson, J. F., Weiss, D. A. et al. Genetic evidence for different male and female roles during cultural transitions in the British Isles. *PNAS* 2001; **98**: 5078–83. See also, Capelli, C., Redhead, N. et al. A Y chromosome census of the British Isles. *Curr Biol* 2003; **13**: 979–84.

Venter and Watson more similar to Korean genome: Barbujani, G. et al. Human races. *Curr Biol* 2013; **23**(5): R185–R187. See also, Ahn, S.-M., Kim, T.-H. et al. The first Korean genome sequence and analysis: full genome sequencing for a socio-ethnic group. *Genome Research* 2009; **19**:1622–29.

Öetzi the Iceman: Ermini, L., Olivieri, C. et al. Complete mitochondrial genome sequence of the Tyrolean Iceman. *Curr Biol* 2008; **18**: 1687–93. See also, Keller, A., Graefens, A. et al. New insights into the Tyrolean iceman's origin and phenotype as inferred by whole-genome sequencing. *Nature Communications* 2012: doi: 2-.1038/ncomms1701 | www.nature.com/naturecommunications.

The origins of different sections of the human genome: Pääbo, S.: 186–187.

Chapter 20

Chapter head quote: Venter, J. C.: 110.

For more lay-directed discussion of scientists and ethics involved see Duncan, D. E., 2006.

Adverse drug reactions: Thiesen, S., Conroy, E. J. et al. Incidence, characteristics and risk factors of adverse drug reactions in hospitalized children – a prospective observational cohort study of 6,601 admissions. *BMC Medicine* 2013; **11**: 237. http://www. biomedcentral.com/1741-7015/11/237.

Chen, R., Mias, G. I. et al. Personal Omics profiling reveals dynamic molecular and medical phenotypes. *Cell* 2012; **148**: 1293–307.

Regulation of GM crops: see Wikipedia article, 'Genetically modified organism containment and escape'.

Wild canola found to contain genetically modified genes: Ibid.

We are still evolving: Voight, B. F., Kudaravalli, S. et al. A map of recent positive selection in the human genome. *PLoS Biology* 2006; **4**(3): e72, 0446–58.

Constructing the first artificial genome: Gibson, D. G., Glass, J. I. et al. Creation of a bacterial cell controlled by a chemically synthesized genome. *Science* 2010; **329**: 52–56.

Human embryo deliberately engineered in a scientific experiment: Liang P, Xu Y, Zhang X, et al. CRISPR/Cas9-mediated gene editing in human tripronuclear zygotes. Protein Cell 2015 doi 10.1007/s13238-015-0153-5.

index

The Mysterious World of the Human Genome

Reich, David 268, 269, 276
retrovirus 16, 101, 155
  co-evolution and 158–9
  constant change within human
    genome and 282, 301
  endogenous (ERVs) 162–71, 173,
    212, 222, 236, 282
  epidemic, eastern Australia koala
    161–2, 223
  epigenetic silencing and 163
  exogenous 162, 163, 223–4
  human endogenous (HERVs) *see*
    human endogenous retroviruses
    (HERVs)
  long terminal repeats (LTRs) and
    164, 165, 169, 193, 195, 223
  origins of *Homo sapiens*, distribution
    of used to locate 212, 222–4,
    236, 288, 301
  retroviral legacy role in holobiontic
    evolution of human genome
    164–71, 173, 212, 222
  RNA based genome 159
  symbioses with host 159–62, 222,
    223
  viral *env* gene 164, 165, 166
  viral loci 162, 163, 164, 165–7,
    168, 169
Rhenen, Netherlands 264
Rich, Alex 96
Richard III, King 207–8
Ridley, Matt 97
Riss (Ice Age) 234
RNA (ribonucleic acid) 8
  discovery of 17–18
  epigenetics and 184, 185–96, 236,
    247, 266, 291, 292, 294
  GACU structural chemicals
    (guanine, adenine, cytosine,
    uracil) 17, 94, 95
  lncRNAs (long non-coding RNAs)
    191–3

mRNA (messenger RNA) 98, 131,
  133, 134, 135, 159, 167, 186,
  187, 188, 190, 191, 193
non-coding 173–4, 187–9, 191–6,
  199–200, 236, 266
piRNAs (Piwi-interacting RNAs)
  189
programmed cell death and 186–9
ribose and 94
ribosomal 187
'RNA gene', idea of 188
RNAi (RNA interference) 187–9
role in gene extrapolation to
  proteins first explored 94–5, 96,
  98, 131–5
single-stranded helix 94
siRNAs (small interactive RNA
  molecules) 186–7
tRNA (transport/transfer RNA) 98,
  187
variance in amount of in different
  cells 94
viral genomes and 157, 159, 160,
  173–4, 183–4 *see also* retrovirus
RNA Tie Club 96
Robert Koch Institute, Berlin 15
Roberts, Alice 269–70
Roberts, Richard J. 131, 132, 133, 135
Roche 246
Rockefeller Institute for Medical
  Research, New York 5, 6, 13, 15–16,
  18, 19, 21, 22–3, 24, 25, 30, 31, 33
Roebroeks, Wil 256, 272
Rothberg, Jonathan 246
rubella, vaccinations against 109
Rudbeck Laboratory 167, 168
Russia 238
Rutgers University, New Jersey 5, 6, 32
Rutherford, Lord Ernest 51

San people 215, 290
Sanger Centre, UK 127

Acknowledgements

Many scientific colleagues have contributed to my thoughts on the human genome over the last two decades, and many more members of my audiences in the lectures I have given on various aspects of this fascinating theme. In particular I am indebted to the kindness and generosity of Erik Larsson and his colleagues at Uppsala, and Katerina Douka at the Oxford Radiocarbon Accelerator Unit for reasons that will become obvious in the text. I would, of course, thank my publisher, Myles Archibald at HarperCollins, for the many communications and conversations that led to the scope and format of the book. It's a great pleasure also to acknowledge the help, and practical suggestions, of my agent, Jonathan Pegg, as well as the ever-supportive editor, Julia Koppitz, at HarperCollins. I thank you all with fondness and gratitude for your interest and stimulation.